INVOLUTE SPUR GEARS

DESIGN & LATHE CUTTING

by

Earle Buckingham

Mechanical Engineering Series

Wexford Press
2008

INTRODUCTION

GEARS are, and always have been, vital factors in machinery. They were used in clocks, one of the first mechanisms invented. In fact, a clock is little more than a train of gears. Despite their constant use during several centuries, our knowledge of gears is slight; and we feel our ignorance keenly because today gears have to meet exacting conditions. They have to transmit heavier loads and to run at higher speeds than ever before. This paper is an attempt to present the problem of gearing. The subjects to be discussed are as follows:

The Involute Curve and Its Properties,

The Design of Involute Gear Tooth Profiles,

Methods of Production,

Methods of Testing,

Strength of Gears.

This paper on Involute Spur Gears is necessarily incomplete as each topic touched on requires a volume for itself. If the reader gains from these pages a clearer idea of the problems involved in designing, producing, and testing gears, the paper has achieved its object.

Testing a Large Gear for Concentricity with Plug Gage and Dial Indicator.
(See page 74.)

CONTENTS

CONTENTS—*Continued*

CHAPTER IV

CHAPTER I

The Involute Curve and Its Properties

Involute Curve Formulae

THE involute curve, extensively used as a gear tooth profile, is generated by the end of a line which is unwound from a circle, as illustrated in Fig. 1. The circle from which the line is unwound is commonly known as the base circle. The equation of the involute curve and its derivation is given below:

Let a = radius of base circle;
   b = length of generating line;
   r = length of radius vector;
   θ = vectorial angle;
   a = angle between radius vector and radial line of base circle to point of tangency of the generating line with the base circle.

It is often necessary to use the circular measure of an angle when dealing with the involute curve. The circular measure of an angle is the length of the arc with a radius of 1, which is subtended by the angle. The circular measure of 180° is equal to π, the circular measure of $90° = \frac{\pi}{2}$, of $45° = \frac{\pi}{4}$, etc. In all equations where the symbol of an angle is used without any expressed trigometric function, the circular measure of the angle is employed.

In Fig. 1, the length of the generating line is equal to the length of the arc subtended by the angles θ and a as this line has been unwound from that portion of the circumference of the base circle. Thus we have

$$b = a\ (\theta + a).$$

But b is also the leg of a right angle triangle, from whence we have

$$b = a \tan a,$$

whence

$$a\ (\theta + a) = a \tan a,$$
$$\theta + a = \tan a,$$
$$\theta = \tan a - a.$$

From the same right angle triangle we have

$$r = \frac{a}{\cos a}.$$

These last two equations are the simplest form in which the equations of the involute of a circle can be given. In some cases, however, it is necessary to use the polar equation of this curve. This polar equation must contain only the terms θ, a, and r. Referring to Fig. 1,

$$\tan \alpha = \frac{b}{a},$$

$$b = \sqrt{r^2 - a^2},$$

$$\tan \alpha = \frac{\sqrt{r^2 - a^2}}{a} = \sqrt{\left(\frac{r}{a}\right)^2 - 1}.$$

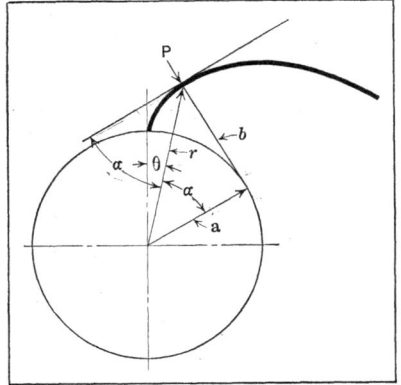

Fig. 1.   The Involute Curve.

α represents the length of an arc with a radius of 1 whose tangent is $\sqrt{\left(\frac{r}{a}\right)^2 - 1}$.   Thus

$$\alpha = \arctan\sqrt{\left(\frac{r}{a}\right)^2 - 1}.$$

The polar equation of the involute curve is therefore,

$$\theta = \sqrt{\left(\frac{r}{a}\right)^2 - 1} - \arctan\sqrt{\left(\frac{r}{a}\right)^2 - 1}.$$

We know that the equation of the tangent to the radius vector is

$$\tan \Psi = r\frac{d\theta}{dr},$$

$$\frac{d\theta}{dr} = \frac{\dfrac{r}{a^2}}{\sqrt{\left(\frac{r}{a}\right)^2 - 1}} - \frac{\dfrac{r}{a^2}}{\dfrac{r^2}{a^2}\sqrt{\left(\frac{r}{a}\right)^2 - 1}}$$

$$\frac{d\theta}{dr} = \frac{r}{a^2}\frac{\left(\frac{r}{a}\right)^2 - 1}{\dfrac{r^2}{a^2}\sqrt{\left(\frac{r}{a}\right)^2 - 1}} = \frac{\sqrt{\left(\frac{r}{a}\right)^2 - 1}}{r}$$

$$r\frac{d\theta}{dr} = \tan \Psi = \sqrt{\left(\frac{r}{a}\right)^2 - 1} = \tan \alpha$$

The radius of curvature at any point is equal to the length of the generating line from its point of tangency to the base circle to the given point. In other words, the radius of curvature at any point is equal to b.

As the angle between the tangent to the involute and the radius vector is equal to a, this tangent will always be perpendicular to the generating line. Thus the generating line is the normal of the involute curve.

Action of Involute As a Cam

A simple conception of the involute curve is that of a uniform rise cam where the rise per revolution along a line tangent to a circle of radius "a" is equal to the circumference of the circle. This is shown in Fig. 2. If this cam is revolving at a uniform rate of speed in the direction shown by the arrow, the cam roll will rise at a uniform rate. If the cam revolves in the reverse direction, the roll will fall accordingly.

This line which is tangent to the base circle is called the line of action or line of contact.

Action of One Involute Against Another Involute

If, instead of acting against a roll, the involute acts against another involute, we have the conditions shown in Fig. 3. The point of contact between two involutes is that point where the tangents to the two curves coincides. The tangents to both involutes are always perpendicular to the generating line. The tangents to two involutes in contact coincide only when the generating line of one involute is a continuation of the generating line of the second involute. Therefore the locus of the points of contact between two involutes is the common tangent to the two base circles.

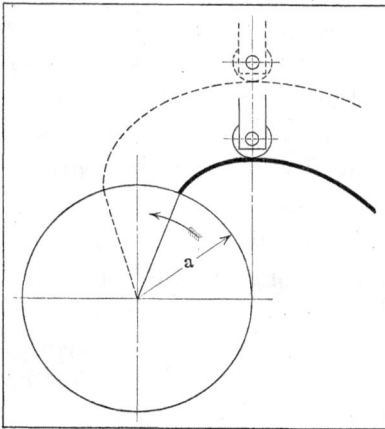

Fig. 2.   Action of Involute as a Cam.

9

When one involute is re-volved at a uniform rate of motion, the length of the line of action, b_1, from its point of tangency with its base circle, a_1, to the involute profile, point P, changes uniformly. If the direction of the rota-tion is in the direction shown by the arrow in Fig. 3, the length of this line increases. At the same time the length of the line of action from the point P to its point of tangency with the base circle a_2, which is b_2, is shortened at a cor-

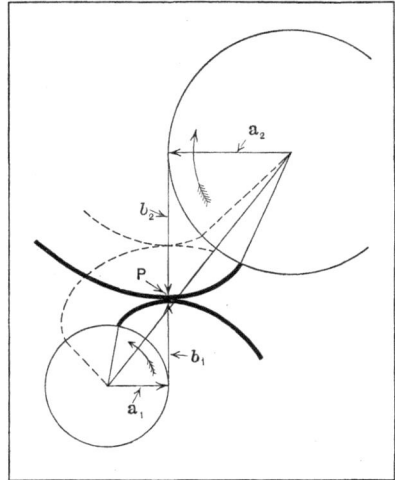

Fig. 3. Action of One Involute Against Another.

respondingly uniform rate because the total length of the common tangent remains constant. This means that the second involute must revolve at a uniform rate in the direction shown by the arrow in Fig. 3.

The relative rate of motion depends only upon the rela tive sizes of the two base circles. No matter what the dis-tance is between the centers of the two base circles, when one involute acts against another, contact between them occurs only along the common tangent to the two base circles and their relative rates of revolution remain the same. If the two base circles are identical in size, this rate is identical for both. If one base circle is double the size of the other, the rate of revolution of the larger involute is one-half that of the smaller. This is because the larger base circle revolves only half as far as the smaller to wind up the length along the line of action that the smaller one has unwound. The conditions are ex-actly the same as though two pulleys were set up connected by a crossed belt. Thus the relative rates of revolution of two involutes which act against each other are in inverse pro-portion to the sizes of their base circles.

When S_1 = rate of revolution of first involute,
S_2 = rate of revolution of second involute,
a_1 = radius of base circle of first involute,
a_2 = radius of base circle of second involute,

$$S_1 : S_2 = a_2 : a_1,$$

$$\text{or } \frac{S_1}{S_2} = \frac{a_2}{a_1}.$$

The relative rates of revolution of the two involutes may be represented by two plain discs which drive each other by friction. Such discs are commonly known as pitch discs, while their diameters are called pitch diameters. An involute has no pitch diameter until it is brought in contact with another involute. This is contrary to a common notion that an involute depends upon a definite pitch diameter. As has been shown, the involute is developed from its base circle only.

In Fig. 4 two involutes are shown in contact at different center distances. The common tangent to the two base circles is the line of action. We have seen before that the radii of the base circles are in inverse proportion to the rates of revolution of the involutes. The radii of two pitch discs which represent the same relative rates of revolution must be directly proportional to the radii of the base circles of their respective involutes.

Thus when
R_1 = radius of pitch circle of first involute,
R_2 = radius of pitch circle of second involute,

$$a_1: R_1 = a_2: R_2$$

$$\text{or } \frac{a_1}{R_1} = \frac{a_2}{R_2}.$$

In Fig. 4 we have two similar right angle triangles in which R_1 is the hypothenuse of one, while R_2 is the hypothenuse of the other. As a_1 is a leg of one and a_2 is the corresponding leg of the other,

$$a_1 : R_1 = a_2 : R_2.$$

The intersection of the common tangent to the two base circles with the common centerline establishes the radii of the two pitch circles.

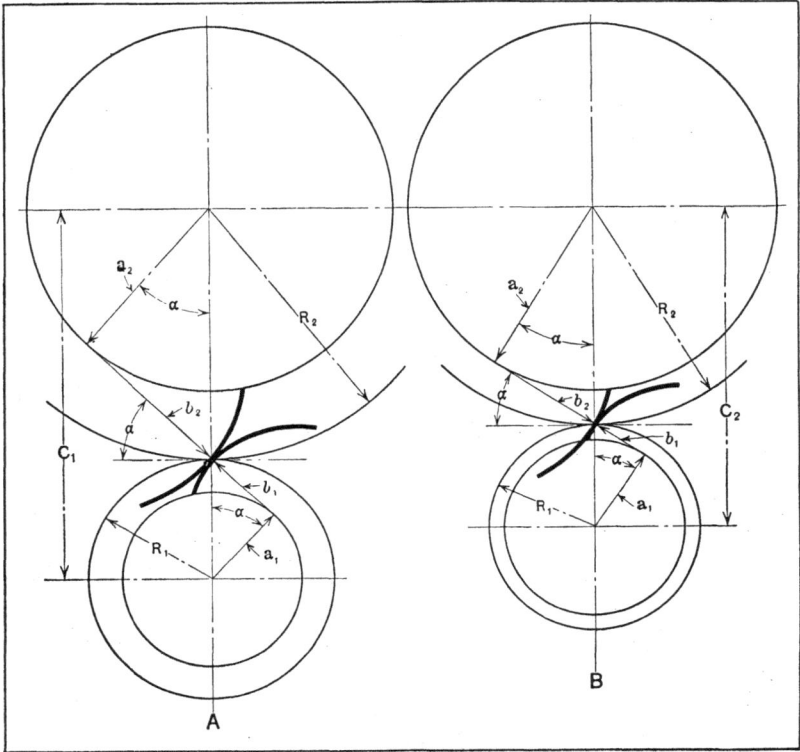

Fig. 4.   Action of Involutes when Set at Different Center Distances.

The angle between the common tangent to the two base circles and a line perpendicular to their common centerline is called the pressure angle.   This angle does not exist until two involutes are brought into contact.   It will be noted that in Fig. 4, although the same involutes are shown at "A" and "B," they have different pitch circles at "A" than at "B" because the distance between the centers of the two base circles is changed.   With the smaller center distances, both the pressure angle and the pitch circles are smaller.

Thus both the pitch circles and the pressure angle of a pair of involutes depend solely upon the sizes of the base circles and the distance between their centers.

When C = center distance,

$$C = R_1 + R_2,$$

$$R_1 = \frac{a_1 R_2}{a_2},$$

$$C = \frac{a_1 R_2}{a_2} + R_2 = R_2\left(\frac{a_1 + a_2}{a_2}\right),$$

Whence

$$R_2 = \frac{a_2 C}{a_1 + a_2},$$

In like manner

$$R_1 = \frac{a_1 C}{a_1 + a_2}.$$

Referring again to Fig. 4, when

a = pressure angle of the pair of involutes,

$$\cos a = \frac{a_1 + a_2}{C}.$$

Action of An Involute Against A Straight Line

When an involute acts against a straight line, we have the conditions shown in Fig. 5. The straight line is tangent to the involute and is always perpendicular to its line of action. When it is constrained to move only in the direction of the line of action, it will be moved at a corresponding and uniform rate if the involute is revolved at a uniform rate. The distance this line moves depends upon the size of the base circle of the involute. Thus for one complete revolution of the involute the line moves along the line of action a distance equal to the circumference of the base circle.

We will now consider the motion of this straight line when it is constrained so that it can move only in the direction of the line $A-A$. If we designate the distance which the line travels along the line $A-A$ as D_1 and the distance the line has moved along the

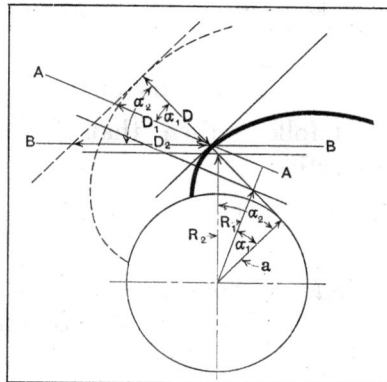

Fig. 5.   Action of Involute Against a Straight Line.

13

line of action as D, and the angle between the line of action and the line $A-A$ as α_1, we have the following relationship:

$$D_1 = \frac{D}{\cos \alpha_1}.$$

As the value of D changes uniformly, and the value of α_1 is constant, the value of D_1 changes uniformly. As $\cos \alpha_1$ can never be greater than 1, the value of D_1 will never be smaller than D. Therefore, when the line is constrained so that it moves only in the direction of the line $A-A$, the distance it travels along this line will be greater than the distance along the line of action, but its rate of motion will be uniform as long as the rate of motion of the involute is uniform.

If the involute should make one complete revolution, the value of D would become $2\pi a$. The value of D_1 would then be $\frac{2\pi a}{\cos \alpha_1}$. This also represents the circumference of a pitch disc which runs with a straight edge parallel to the line $A-A$. The radius of the pitch circle thus becomes $\frac{a}{\cos \alpha_1}$. Referring to Fig. 5, and designating the radius of this pitch line by R_1, this radius would be determined by the intersection of a radial line of the base circle perpendicular to $A-A$ with the line of action.

In like manner, when the line against which the involute is acting is constrained so that it moves only along the line $B-B$, the distance R_2 becomes the radius of the pitch circle of the involute and the angle α_2 becomes its pressure angle. The motion along the line $B-B$ is also uniform when the rate of movement of the involute is uniform. The distance $D_2 = \frac{D}{\cos \alpha_2}$ and $R_2 = \frac{a}{\cos \alpha_2}$.

Summary of Involute Curve Properties

It follows then, that the involute curve has the following properties:

First: The involute is dependent only upon the size of its base circle.

Second: If one involute, rotating at a uniform rate of motion, acts against another involute, it will transmit a uniform angular motion to the second involute, regardless of the distance between the centers of the two base circles.

14

Third: The rate of motion transmitted from one involute to another depends only upon the relative sizes of the base circles of the two involutes. This rate of angular motion is in inverse proportion to the sizes of the two base circles.

Fourth: The common tangent to the two base circles is the line of action. In other words, the two involutes are in contact only along this common tangent.

Fifth: The intersection of the common tangent to the two base circles with their common center line determines the pitch line of the two involutes. No involute has a pitch line until it is brought in contact with another, or a straight line constrained to move in a defined fixed direction.

Sixth: The pitch diameters of two involutes acting together are directly proportional to the diameters of their base circles.

Seventh: The pressure angle of two involutes acting together is the angle between the common tangent to the two base circles and a line perpendicular to their common center line. No involute has a pressure angle until it is brought in contact with another involute or straight line constrained to move in a definite fixed direction.

Eighth: The pressure angle of an involute acting against a straight line constrained to move in a fixed direction is the angle between the line of action and a line representing the direction in which the straight line can move.

Ninth: The pitch radius of an involute acting against a straight line constrained to move in a fixed direction is the distance along a radial line of the base circle of the involute, which is perpendicular to the direction in which the straight line can move, to its intersection with the line of action.

15

Use of Involute Form For Gear Tooth Profiles

When the involute form is used as a gear tooth profile, several involute curves are developed from the same base circle to form the profiles of the several teeth. As the gear tooth profile is symmetrical, for the present we shall consider but one side of the teeth.

Fig. 6 shows the development of one side of the several teeth. The distance between

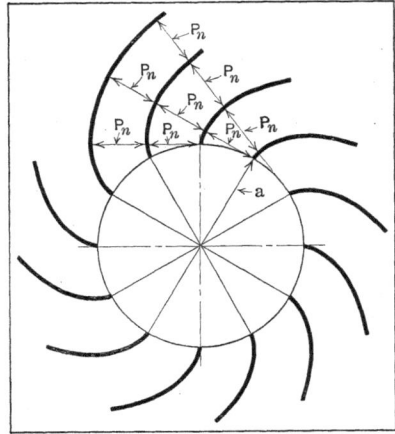

Fig. 6.   Equally Spaced Involutes Developed from Common Base Circle.

these involute curves, measured along any line tangent to the base circle, is always the same. This distance is equal to the length of the arc of the base circle between the origins of two successive involutes, and is the normal pitch of the gear.

Thus when

P_n = normal pitch,

a = radius of base circle,

N = number of teeth in gear,

$$P_n = \frac{2\pi a}{N} .$$

In a pair of mating gears the normal pitch must be identical in order to secure smooth action.

Duration of Contact

One of the important factors in the design of gears which are to transmit power is that the involute profiles must be so selected that the second pair of mating teeth will be in contact before the first pair are out of contact. The proper amount of overlap depends upon several conditions. If the form of the tooth is rugged and the load when applied towards the outer end of the tooth causes practically no deflection, the amount of overlap may be small with satisfactory results.

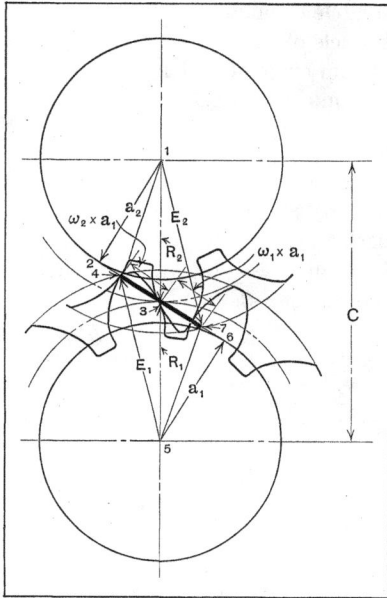

Fig. 7.   Duration of Contact.

The arc of action refers to the arc through which one tooth travels from the time it makes contact with its mating tooth until it ceases to be in contact. The number of teeth in contact is the quotient of the arc of contact divided by the arc between two successive teeth on the gear. Thus if an overlap of 0.6 exists, the number of teeth in contact is 1.6.

In Fig. 7, that part of the line of action which is intercepted by the two outside circles of a pair of gears, shown as a heavy line, is the length of the arc of action with a radius equal to the radius of the base circle. This length divided by the normal pitch, which is the length of an arc of the same radius between two successive teeth on the gear, gives the number of teeth in contact.

The arc of action is often divided into the arc of approach and the arc of recess. The arc of approach is the arc through which the tooth moves from the time it comes into contact until the point of contact reaches the pitch line. The arc of recess is the arc through which the tooth moves from the time when the point of contact is at the pitch line until the tooth ceases to be in contact.

Referring to Fig. 7, let

a = pressure angle,
ω_1 = arc of approach,
ω_2 = arc of recess,
N_1 = number of teeth in pinion,
N_2 = number of teeth in gear,
C = center distance,

17

$R_1 = $ radius of pitch circle of pinion,
$R_2 = $ radius of pitch circle of gear,
$E_1 = $ radius of addendum circle of pinion,
$E_2 = $ radius of addendum circle of gear,
$P_n = $ normal pitch.

Simple formulae can be derived for the values of ω_1 and ω_2 and the number of teeth in contact by solving various right angle triangles. The angle of approach, ω_1, in circular measure, is found by dividing the length of the line 3–7 by a_1. This length is equal to the length 2–7 minus the length 2–3.

Length of line 2–3 $= R_2 \sin \alpha$,
Length of line 2–7 $= \sqrt{(E_2)^2 - (a_2)^2}$.

Whence $$\omega_1 = \frac{\sqrt{(E_2)^2 - (a_2)^2} - R_2 \sin \alpha}{a_1}.$$

The angle of recess, ω_2, is found in similar manner by dividing the length of the line 3–4 by a_1. This length is equal to the length 4–6 minus the length 3–6.

Length of line 3–6 $= R_1 \sin \alpha$.
Length of line 4–6 $= \sqrt{(E_1)^2 - (a_1)^2}$.

$$\omega_2 = \frac{\sqrt{(E_1)^2 - (a_1)^2} - R_1 \sin \alpha}{a_1}.$$

The number of teeth in contact is found by dividing the length of the line 4–7 by the normal pitch P_n. The length of the line 4–7 is equal to the sum of 3–4 and 3–7. The normal pitch P_n is equal to $\frac{2\pi a_1}{N_1}$ or $\frac{2\pi a_2}{N_2}$, as shown previously.

Length of line 4–7 $= \sqrt{(E_2)^2 - (a_2)^2} + \sqrt{(E_1)^2 - (a_1)^2} - (R_1 + R_2) \sin \alpha$.

$$R_1 + R_2 = C.$$

Whence

Number of teeth in contact $=$

$$\frac{\sqrt{(E_2)^2 - (a_2)^2} + \sqrt{(E_1)^2 - (a_1)^2} - C \sin \alpha}{P_n}.$$

Sometimes the tooth form of a gear extends below the base circle. No involute action, however, can take place below the base circle. Thus if the value of $\sqrt{(E_2)^2 - (a_2)^2}$ or $\sqrt{(E_1)^2 - (a_1)^2}$, is greater than $C \sin \alpha$, which is the total length of the line of action, the value of $C \sin \alpha$ must be substituted in place of the greater value because the line of actual contact does not extend beyond the point of tangency to either base circle.

18

Sliding and Rolling Contact

As pointed out before, the length of the generating line which is unwrapped from the base circle is the radius of curvature of the involute at any point. Fig. 8 shows the position of this generating line at equal angular distances. At the origin (1) the length of the generating line is zero. At (2) it is infinitely longer. At (3) it is twice the length

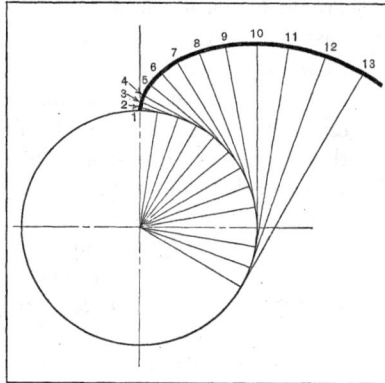

Fig. 8.   Lengths of Involute Profile.

at (2).   At (4) it is one and one-half times the length at (3). At (5) it is one and one-third times the length at (4), and so on.   The radius of curvature thus changes rapidly in proportionate length near the base circle and more slowly as the curve departs further from the base circle.   In other words, the form near the base circle is very sensitive—that is, it has a small and rapidly changing radius of curvature—but becomes less sensitive the further it departs from the base circle.

It will also be noted that the length of the curve 1–2 is much less than the length of 2–3; that 2–3 is smaller than 3–4, etc.   Thus whether the involute is acting as a cam or is acting against another involute, the length of the curve that must pass through the line of action for any series of equal angular movements changes constantly.   The nearer the active part of the profile is to the base circle, the shorter the length of profile.

Thus when two involutes are acting against each other, a combined rolling and sliding action takes place between them because of the varying lengths of equal angular increments on the profiles.

Fig. 9 shows two equal involutes with the generating lines shown at equal angular distances. The part of the profile 1–2 on one involute comes into contact with the profile 7–8 on the other.

19

Profile 1–2 is much nearer its base circle than 7–8 and is therefore much shorter. The two profiles must slide against each other a distance equal to their difference in length to make up this difference. Profile 2–3 is longer than 1–2, while profile 8–9 is shorter than 7–8. The length 2–3 is still much shorter than 8–9, but the amount of sliding will not be as great as with the previous pair of mating profiles. Spaces 3–4 and 9–10 are more nearly equal in length, 3–4 being the shorter, so that still less sliding occurs. The profiles 4–5 and 10–11 are almost equal in length, but the length of the profile 4–5 on the first involute

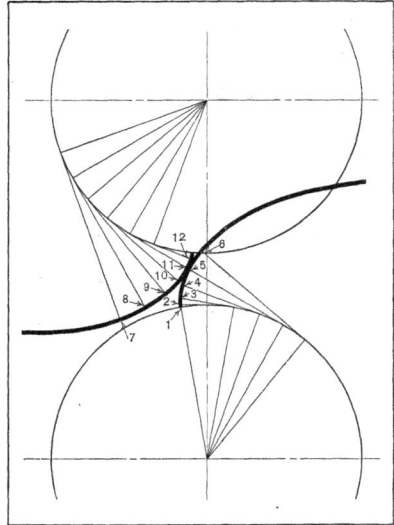

Fig. 9.   Sliding Action Between Two Involutes.

is now slightly longer than its mating portion of the profile on the second involute. Thus, the slight amount of sliding that occurs now takes place in an opposite direction. The remaining sections of the profile of the first involute become increasingly longer, while those on the second involute become smaller, so that the amount of sliding increases again.

It will be seen that the rate of sliding between two involutes acting against each other is constantly varying. The rate of sliding decreases to zero, changes its direction, and increases again. The actual amount of sliding is the same on both profiles, but its rate or velocity is different.

This condition of sliding can be illustrated in a simple manner by considering the action between two discs of equal diameter. When these discs are rolled together at the same rate so that each makes a complete revolution in the same time, pure rolling action occurs. This represents the action of two involutes in contact on the pitch line. When one of the discs is held stationary while the other revolves, sliding results. The amount of sliding is the same on both discs, but it is concentrated at one point on the stationary disc while distributed over the whole surface of the revolving disc. The sliding action on the stationary disc represents the sliding action at the base circle of an involute gear tooth while that on the revolving disc represents the sliding on that part of the addendum of the mating tooth which makes contact at the base circle. Again, when both discs are rolled together but each rotates at a different rate of speed, sliding develops. The amount of sliding is the same on both discs but it is distributed over a larger part of the circumference of the rapidly revolving disc than of the slower one. The sliding action on the faster disc represents the sliding action on the addendum of an involute gear tooth, while that on the slower disc represents the sliding on the dedendum of the tooth.

Formulae for computing the rate of sliding, or specific sliding, as it is called, are derived as follows: The value of the specific sliding multiplied by the length of the profile gives the actual amount of sliding. Or, in other words, the specific sliding represents the quotient of the actual amount of sliding divided by the length of the profile on which this sliding occurs. Referring to Fig. 10, let

N_1 = number of teeth in pinion,
N_2 = number of teeth in gear,
a_1 = radius of base circle of pinion,
a_2 = radius of base circle of gear,
b_1 = length of generating line of pinion to point of contact,
b_2 = length of generating line of gear to point of contact,
r_1 = radius on pinion to point of contact,
r_2 = radius on gear to point of contact,
C = center distance,
a = pressure angle.

We may consider any infinitely small portion of the involute profile as an arc of a circle with a radius equal to the length of the generating line to that point. The lengths of two such infinitely small arcs of the same angular magnitude will be proportional to the lengths of their radii. But the angular magnitude of two such infinitely small arcs which form the point of contact of two involutes is inversely proportional

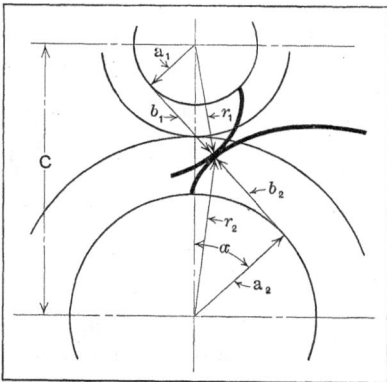

Fig. 10. Diagram for Determining Formulae for Rate of Sliding.

to the sizes of their respective base circles, or the respective number of teeth in the gears. Therefore, if the ratio of the lengths of the radii of the two infinitely small arcs in contact are directly proportional to the number of teeth in the two gears, pure rolling will result. If these lengths are not thus proportional sliding will result. The rate of sliding will be equal to the difference between the relative lengths of these two infinitely small arcs divided by the relative length of one of them. Thus, we may write as the formula for determining the rate of sliding, or specific sliding, the following:

$$\text{Specific sliding on pinion at point } r_1 = \frac{b_1 N_2 - b_2 N_1}{b_1 N_2}.$$

$$\text{Specific sliding on gear at point } r_2 = \frac{b_2 N_1 - b_1 N_2}{b_2 N_1}.$$

The total length of the line of action, or common tangent to the two base circles $= b_1 + b_2 = C \sin a.$

$$b_1 = \sqrt{(r_1)^2 - (a_1)^2} \text{ and } b_2 = C \sin a - b_1.$$

When the portion of the involute of the pinion at the base circle is in contact, we have the following:

$$b_1 = 0.$$
$$r_1 = a.$$
$$\text{Specific Sliding} = \frac{0 - b_2 N_1}{0} = -\infty.$$

In like manner the specific sliding at the base circle of the gear will be found equal to minus infinity.

When the portion of the involute of the pinion at the pitch circle is in contact, we have the following:

$$b_1 = a_1 \tan a,$$
$$b_2 = a_2 \tan a.$$

$$\text{Specific sliding} = \frac{a_1 N_2 \tan a - a_2 N_1 \tan a}{a_1 \tan a N_2} = \frac{a_1 N_2 - a_2 N_1}{a_1 N_2}.$$

But

$$a_1 : a_2 = N_1 : N_2.$$

Whence

$$a_1 N_2 = a_2 N_1.$$

Substituting, we have,

$$\text{Specific Sliding} = \frac{0}{a_1 N_2} = 0.$$

Thus, pure rolling action occurs on the pitch line of the pinion. In like manner it will be found that pure rolling occurs on the pitch line of the gear. The portions of both involutes at the pitch line make contact with each other, thus pure rolling occurs only at the pitch line.

A rack may be considered as an involute curve with an infinitely large base circle. The lengths of the portions of the profile corresponding to equal angular movements of the pinion will be equal. In order to derive formulae for the specific sliding of a rack meshing with a pinion, we will start from the pitch line, where we know that pure rolling occurs. Thus, the length of the infinitely small portion of the profile of the rack, that corresponds to the mating portion of the profile of the pinion, is equal to the length of the profile of the pinion at

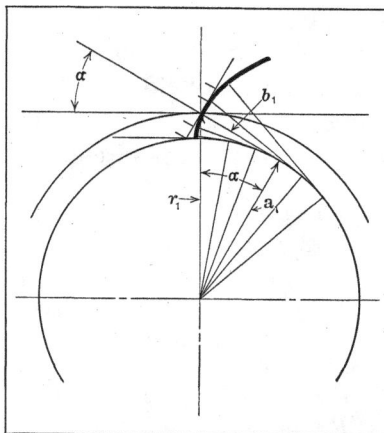

Fig. 11. Diagram for Determining Formulae for Sliding on Rack.

that point. The corresponding radius of curvature on the pinion profile at that point is equal to $a_1 \tan a$. (See Fig. 11.) Thus, we can write the following formulae for the specific sliding of a rack and pinion:

$$\text{Specific sliding on pinion} = \frac{b_1 - a_1 \tan a}{b_1}.$$

$$\text{Specific sliding on rack} = \frac{a_1 \tan a - b_1}{a_1 \tan a}.$$

Undercutting of Involute

As the involute curve stops at the base circle, no involute action can take place below it. If a straight-sided rack with sharp corners acts against the involute, and these extend too far below the base circle, interference develops unless the tooth is undercut as shown in Fig. 12. The looped curve shows the path of the sharp corner of the rack tooth. In order to avoid this interference, the corner of the rack can extend below the

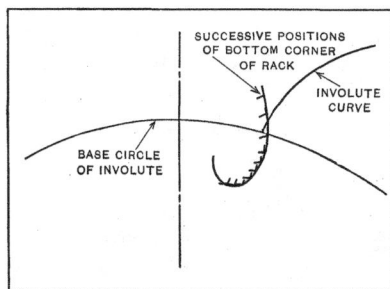

Fig. 12. Undercutting of Involute by a Rack.

base circle only a limited amount. Its bottom edge must not reach below the line where the line of action is tangent to the base circle.

In Fig. 13, let

A = minimum distance between bottom of straight-sided rack tooth with sharp corners and center of base circle.

a = radius of base circle;

R = radius of pitch circle of involute;

α = pressure angle;

$A = \text{a} \cos \alpha = R \cos^2 \alpha.$

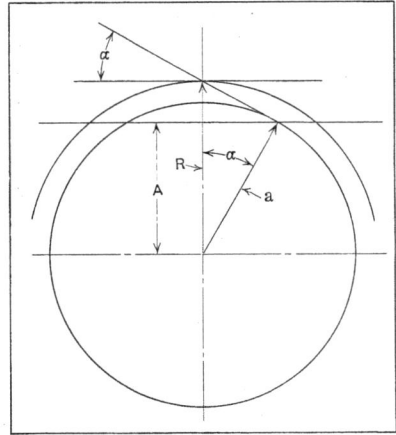

Fig. 13.   Undercut Limit for Rack.

In similar manner, if two involutes are acting against each other, their outside diameters must not reach beyond the point of tangency of the line of action and the base circle or a similar interference will develop.

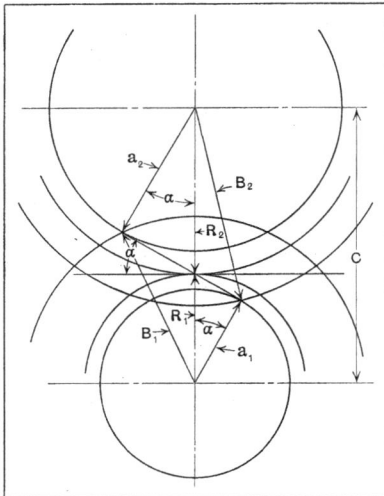

Fig. 14.   Undercut Limit for Two Gears.

In Fig. 14, let

B_1 = radius of maximum addendum circle of pinion without interference;

B_2 = radius of maximum addendum circle of gear without interference;

C = center distance;

a_1 = radius of base circle of pinion;

a_2 = radius of base circle of gear;

R_1 = radius of pitch circle of pinion;

R_2 = radius of pitch circle of gear;

α = pressure angle;

$B_1 = \sqrt{(a_1)^2 + (C \sin \alpha)^2};$

$B_2 = \sqrt{(a_2)^2 + (C \sin \alpha)^2}.$

Tooth and Bearing Pressures

The tooth pressure of a gear is the pressure exerted in the direction normal to the tooth profile. This pressure on involute gears is exerted along the line of action. Thus both the direction and amount of pressure is constant. This is one of the features which make the involute form of value as a tooth profile.

Referring to Fig. 15, let

T = tangential force action on pitch line of involute;

T_1 = tooth pressure;

a = pressure angle;

$$T_1 = \frac{T}{\cos a}.$$

The pressure on the bearings of the gears will be found in a similar manner.

Referring again to Fig. 15, let

T_2 = force tending to separate axes of gears;

$T_2 = T \tan a$.

This force which tends to separate the axes of the gears is often confused with the total bearing pressure. As a matter of fact, it is but one moment of that force, the other moment being equal to T. Thus the total bearing pressure becomes

$$\sqrt{T^2 + T_2{}^2} = T_1 = \frac{T}{\cos a}.$$

The total bearing pressure and the tooth pressure are thus identical.

A few computations will show that there is only a small difference in tooth pressure or bearing pressure between pressure angles of $14\frac{1}{2}$ degrees and 25 degrees; these angles cover the range most often used.

For example,

Let $T = 1,000$;

$a = 14\frac{1}{2}°$;

$$T_1 = \frac{1,000}{\cos 14\frac{1}{2}°} = 1,035.$$

Let $T = 1,000$;

$a = 25°$;

$$T_1 = \frac{1,000}{\cos 25°} = 1,103.$$

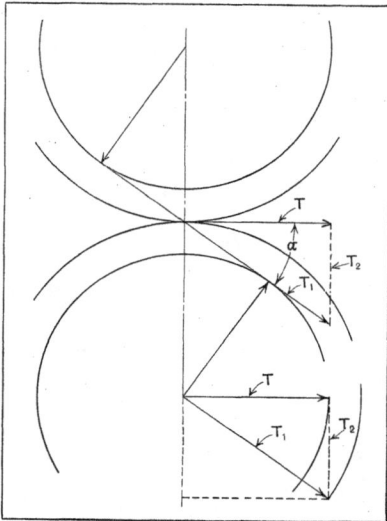

Fig. 15.   Diagram of Tooth and Bearing Pressures.

The pressures are increased less than 7 per cent when the pressure angles are increased from $14\frac{1}{2}°$ to 25°. The greatest effect on the bearing pressure caused by a change in pressure angle is a change in direction.

CHAPTER II

Design of Involute Gear Tooth Profiles

SOME of the characteristics of the involute curve have been discussed. We will now consider how to utilize these characteristics to the best advantage. Many factors must be taken into account in the design of the tooth profiles. In some cases, to obtain the maximum benefit of one factor may involve the loss of another valuable property. As with all engineering problems, we must weigh the importance of one factor against another and select the combination which offers the best compromise between conflicting elements.

One word of caution at the outset: No improvement in form alone will outweigh the importance of careful and painstaking workmanship in the production of gears themselves. In addition, the design of the mechanism in which the gears are used must provide a sufficiently rigid support to hold the gears in alignment under their working loads.

Essential Requirements of Satisfactory Gears

Several of the essential requirements of quiet-running and long-lived gears are given below. It would be well to state at once that no metal gears are absolutely noiseless. Quietness is a relative term. The most accurate gears when running under load at any appreciable speed will develop a certain amount of noise. When such gears are enclosed in cases, as gears often are, this noise is often almost entirely muffled. There are many gear noises. They vary from the unobtrusive hum of the better gears to rumbles and screams of varying pitches and intensities. The reasons for all the noises are not as yet fully known.

The question is not why gears are noisy, but why they are ever quiet. Certain defects in the gears, however, develop certain conditions of noise; great care is required to reduce these defects to a minimum.

We have already noted that the mountings for gears should be rigid and accurately aligned. Unsatisfactory gear conditions are not always due to improper gears. When the mounting is at fault, even perfect gears will be unsatisfactory.

The height of the teeth and the normal pitch of the involute profiles for any gear combination must be so selected that a sufficient overlap is secured. As noted before, if the form of the tooth is strong, the amount of overlap may be small with satisfactory results. Such gears may have but 1.15 teeth in contact. This is a limiting condition. Good engineering requires that a design be kept a safe distance inside limiting conditions. For general practice it is well to keep the minimum number of teeth in contact from 1.25 to 1.4.

The requirements of good workmanship are essential regardless of the design of the tooth profiles. These requirements include smooth surfaces, uniform spacing and concentric profiles.

The primary purpose of gears is to transmit uniform motion. It is self-evident that the involute form of the gear tooth profile must be accurate to achieve this result.

The portion of the involute used for tooth profiles should be chosen so as to avoid excessive sliding. Excessive sliding tends to noise and wear.

The length of the active profiles—that is, the length of the profile which actually comes in contact with its mating tooth—should be as long as possible, in order to distribute the wear over a greater surface.

The radius of curvature of the profile should be as large as other conditions permit. This secures a less sensitive profile, which will resist crushing under load, and gives a stronger and longer lived tooth.

The shape of the tooth should be as strong and rugged as possible. To obtain this strength, the tooth ought to be thickest at the bottom. This will occur automatically if the involute curve forms the entire profile of the tooth. In no case should the working depth of tooth go below the point where interference or undercutting begins.

The pressure angles used for involute gears are generally between the angles of 14½ degrees to 25 degrees. It makes relatively little difference as regards quietness, wear, etc., which angle within this range is used. Since the tooth and bearing pressures increase slightly as the pressure angle is increased, it is well to keep the pressure angles as low as the other more important conditions permit.

Gear Tooth Parts

The several parts of gear teeth have been given definite names. Among them are the following:

DIAMETRAL PITCH. Sometimes the pitch diameter of the gear makes a convenient diameter from which to calculate center distances and tooth proportions, especially when a constant pressure angle is used. The diametral pitch represents the number of teeth to one inch of diameter of the pitch circle. If a gear of 6 diametral pitch has 12 teeth, its pitch diameter equals 12/6 inches, or 2 inches.

CIRCULAR PITCH. The circular pitch is the length of an arc of the pitch circle that corresponds to one tooth interval. It is equal to the circumference of the pitch circle divided by the number of teeth.

MODULE. The module is the reciprocal of the diametral pitch. It represents the size of the pitch diameter per tooth in the gear. The module of a 6 diametral pitch gear equals 1/6. A gear of 1/6 module with 12 teeth would have a pitch diameter of 12 x 1/6 or 2 inches.

ADDENDUM. The addendum of a gear is the height of the tooth outside the pitch circle. This is illustrated in Fig. 16.

DEDENDUM. The dedendum is the depth of the working face of the gear tooth below the pitch circle. The dedendum circle is the circle that is tangent to the outside diameter of its mating gear. See Fig. 16.

CLEARANCE. The clearance is the space below the dedendum which is provided for rounding the bottom of the teeth and providing clearance for the outside diameter of the mating gear.

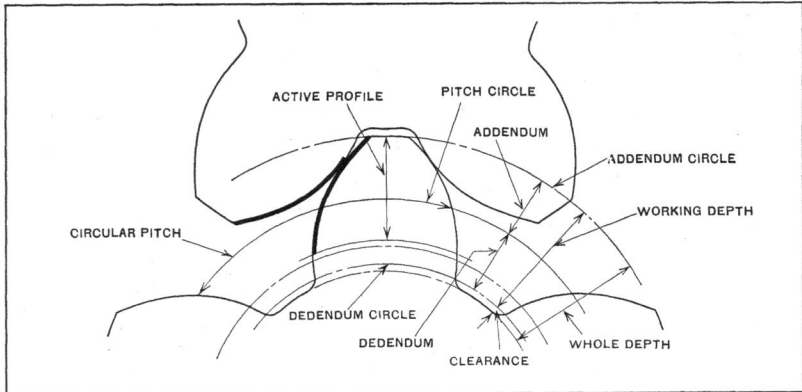

Fig. 16. Names of Tooth Parts.

WORKING DEPTH. The working depth of a tooth is equal to the sum of the addendum and dedundum. It is the distance between the outside diameter and the dedendum circle.

WHOLE DEPTH. The whole depth of the tooth is equal to the sum of the working depth and the clearance. It is the total depth of the space between the teeth.

ACTIVE PROFILE. The active profile is that part of the involute profile which actually comes in contact with the profile of its mating gear along the line of action. In Fig. 16 the active profiles are shown in heavy lines.

Interchangeable Involute Gear Tooth Forms

At the time the involute curve was first adopted as a gear tooth profile, the only method in general use for cutting gears was the process of milling with formed cutters. At first each designer made his gear tooth forms to suit his own needs. This led to a wide variety of tooth forms and a large number of form cutters.

At that time, quantity manufacture as we know it today did not exist. Mechanical appliances of all kinds were made in small numbers. In order to reduce the number of special form cutters required, a system of standard interchangeable gear tooth forms was developed; strictly a measure of economy. At that time a very large proportion of gears had their teeth cast on them.   These did not prove satisfactory for all purposes, and the cut gear was a marked improvement.

Standard interchangeable tooth forms are based on the proposition that all gears of the same diametral pitch with any number of teeth will run together at proportional center distances.   This reduces the calculations to a minimum.

These interchangeable involute gear tooth forms involve a constant pressure angle and a constant addendum and dedendum for all tooth numbers, the addendum and dedendum being equal.

In order to carry out this system, the tooth forms require more or less modification to avoid interference.   In some cases almost all of the involute form is lost.   As mentioned before, the designing of gear tooth forms is the same as any other engineering problem.   The final result is a compromise between conflicting elements.   In this case, several favorable gear conditions have been sacrificed in the interest of economy of tools and simplicity of calculation.   At the time when the involute system of gear tooth forms was coming into general use, gears with small numbers of teeth were seldom employed, so that in the majority of cases fairly good tooth action was secured.

The advent of the automobile, with the demand for reduction in weight, led to the use of gears with smaller numbers of teeth.   This led to the development of another system of interchangeable involute gear forms of considerably greater pressure angle.   This system has stronger tooth forms than the first, particularly on gears with small numbers of teeth, and gives better tooth action on the small gears than the original system.   Both systems are widely used today.   With careful attention to the essential details of producing them, both are giving excellent results.

Manufacturing conditions today are entirely changed from what they were at the time the interchangeable involute system was developed.   Considerations which then were of paramount importance have little weight now.   Then, the component part of a mechanism which was made in sufficient quantities to justify the expense of special form cutters was a great exception. Today such component parts are common. Furthermore, methods of generating the gear tooth forms instead of milling them with form cutters are in general use. A generating process offers many opportunities of producing improved gear forms without the expense of special tools.

It is true that to secure such improved forms, interchangeability as generally understood will be lost. Nevertheless, all gears which have the same normal pitch will run together. Only proportional center distances are lost when a train of gears is employed. When pairs of gears are involved, the commonly known standard center distances are as good as any. Furthermore, if a train of gears of improved form is required to replace an old train which runs at standard center distances, the desired results can often be obtained by adding or subtracting a tooth from an idler gear in the train, thus increasing or reducing the diameter of the base circle of that gear so that a more effective pressure angle can be used. Thus with a little study, improved gear forms can be used not only in new construction, but also as replacements in existing equipment.

Great stress is often laid upon the importance of keeping the tooth forms interchangeable at standard center distances. A little study will show that except for economy in tools where form milling methods are employed, on general jobbing work or for the convenience of the designer who desires to use only the simplest of arithmetical formulae, this factor of interchangeability has little or no importance.

A large proportion of gear drives consist only of pairs of gears or series of pairs. Universal interchangeability has no importance here. Furthermore, a large percentage of gear blanks can only be used in the particular place for which they are designed. This is evident from a study of Fig. 17, which shows some representative gears used in machine tool construction. Universal interchangeability has no importance here.

The requirements and demands which gears today are called on to meet are no mean ones. They have grown with the rapid development of the manufacturing industries, to such an extent that only the most accurately made gears can meet them. This explains why the manufacture of gears of all kinds is being taken care of more and more by those establishments which make a specialty of it and which are thus enabled to devote all their attention and experience to it.

The standard spur gears have been unable to meet some of the requirements. This has led to the introduction of helical and herringbone gears. Incidentally, the center distances at which the helical gears operate are seldom standard. Furthermore, each pair or train is complete in itself and the tooth forms will not interchange with those of other pairs or trains. Yet, although they are in extensive use, the loss of interchangeability has made no trouble.

Helical and herringbone gears are extensively used when high speeds are required. Spur gears of standard form are seldom used where the pitch line velocities exceed 2400 feet per minute. Yet spur gears of involute form, designed to secure the most favorable involute action, are being used to transmit power when running at a pitch line velocity of over 10,000 feet per minute.

In principle, standardization is a desirable goal. But if any standardization acts as a bar to progress in the art, and imposes improper functional conditions, it should be abandoned. No standard should ever be used unless it meets the peculiar requirements of functioning and service in each individual case as well as, or better than any other construction.

Fig. 17. Representative Gears Used in Machine Tool Construction.

The terms "functioning" and "service" are used here in their broadest sense. Service includes the furnishing to the customer of satisfactory mechanisms at the lowest possible cost. The requirements which gears have to meet are not always severe, particularly when their speed is low. In such cases, certain refinements are unnecessary; the cheapest gear to produce is the best gear. As the speed increases, however, or as the amount of power to be transmitted increases, or when maximum strength with minimum weight is essential, no possible refinement can be ignored.

The limitations of the standardized interchangeable gear tooth systems have long been known. Many suggestions have been made for their improvement. In some cases, increased pressure angles have been employed, constant addendum and dedendum being still retained. In other cases, the pinion has an increased addendum, while the gear has a correspondingly increased dedendum.

In order to exploit the involute curve to the greatest extent, the pressure angles and the tooth proportions must be variable. If this required special tools for every gear, the cost of producing such gears would be prohibitive. But as a matter of fact, when a generating process is used, a few standard cutters are sufficient to meet almost every condition, hence the cost of production is no greater.

Representative Standard Tooth Forms

Brown & Sharpe Standard

Among the first standard involute gear tooth forms to be developed was the Brown & Sharpe standard, which maintains a constant pressure angle of 14½ degrees. The tooth forms of these gears are interchangeable. The system is based on two 12-tooth pinions as the smallest pair of gears. The same pinion runs with gears of all tooth numbers, including racks.

The proportions of the teeth are constant throughout all pitches. The addendum and dedendum are each equal to the module. The clearance is equal to one-twentieth of the circular pitch. The thickness of the tooth and space on the pitch line are equal. This thickness is equal to one-half the circular pitch.

Fellow's Standard

The introduction of the automobile created a demand for a tooth form better adapted to gears with small numbers of teeth than the standard 14½ degree form. The Fellow' 20-degree stub tooth form was devised to meet this need. This is an interchangeable tooth, similar to the 14½-degree standard form, but with an increased pressure angle and a shorter tooth. The pitches are denoted by fractions, the numerator being the diametral pitch while the denominator indicates the working depth of the teeth; these working depths are the same as for the standard 14½-degree gear with a diametral pitch equal to the denominator. For example, a 6–8 pitch stub-tooth has the same circular pitch as a 6-pitch standard 14½-degree gear, and the same working depth of tooth as an 8-pitch standard 14½-degree gear. The clearance is equal to one-eighth of the working depth of tooth.

Maag Standard

Both the foregoing tooth forms require some modification of the involute on the smaller tooth numbers to prevent interference. To overcome the need of modification and to utilize the advantage of the involute, the Maag standard gears were developed. These are true involute gears, free from all the restrictive standardization which exists in most standard gear forms. The pressure angles and the proportions of the teeth vary with the different combinations of tooth numbers. The tooth forms are designed to give as long an active profile as possible, with a sufficient overlap or number of teeth in contact, and as low a rate of sliding as other conditions permit. The involute form is not modified in any way. The tooth forms, being pure involutes, are very strong. Pairs of gears may be made to run at standard or special center distances. For a train of gears, either the center distances or the tooth numbers of some part of the train must be variable to get the best tooth action. When the center distances and tooth numbers are fixed, Maag gears of true involute form can be made to meet the conditions. The tooth action, however, will not be as good as when one of these two factors is variable.

No constant tooth proportions can be given because they do not exist in the Maag system. The twelve-tooth pinion which runs with another twelve-tooth pinion is different from the twelve-tooth pinion which runs with a thirteen-, fourteen-, fifteen-, twenty-, or sixty-tooth gear, etc. It is impossible to have gear forms of constant proportions with the best involute action.

Comparison of Standard Tooth Forms

The next consideration is a comparison of the action on the involute profiles of a few samples of the three foregoing types of gears. In every case we will take gears of one diametral pitch. In the case of the Fellow's stub-tooth gear, we will use the 8–10 pitch as a representative form and multiply the dimensions by eight to reduce it to one diametral pitch for the purposes of comparison.

Our first example will be a pair of twelve-tooth pinions. These pinions have the following values:

	B. & S.	Fellows	Maag
Number of teeth	12	12	12
Pitch radius	6.000	6.000	6.000
Pressure angle	14° 30′	20°	24° 46′
Radius of addendum circle	7.000	6.800	6.834
Radius of dedendum circle	5.000	5.200	5.166
Clearance	.157	.200	.139
Working depth of tooth	2.000	1.600	1.668
Radius of base circle	5.809	5.638	5.448
Circular pitch	3.1416	3.1416	3.1416

These are shown in Fig. 18. The involute portions of the tooth profiles are shown in double lines, their active profiles

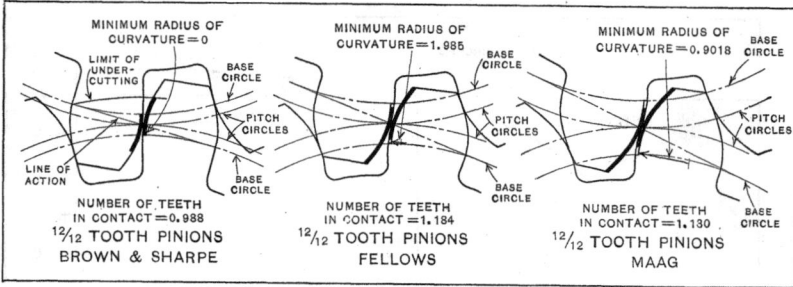

Fig. 18.   Examples of 12/12 Tooth Pinions.

in heavy lines.   It will be noted that the profile of the teeth of the 14½-degree pinions extend for a considerable distance below the limit of undercut.   The profile of the mating gear must be modified to prevent interference with the radial flank. Therefore the involute portion of the tooth profile is much reduced.

A comparison of the three tooth forms shows the following:

The active profile of the Brown & Sharpe standard is the shortest; that of the Fellow's is considerably longer, that of the Maag the longest.

The minimum radius of curvature of the Brown & Sharpe standard is zero, that of the Fellow's is very little more, that of the Maag much greater.

The number of teeth in contact on the Brown & Sharpe pinion equals 0.988; on the Fellow's 1.184; and on the Maag 1.130.

The sliding diagrams of these gears are shown in Fig. 19. Note that the diagram for the Maag gears shows a much flatter curve than either of the others.

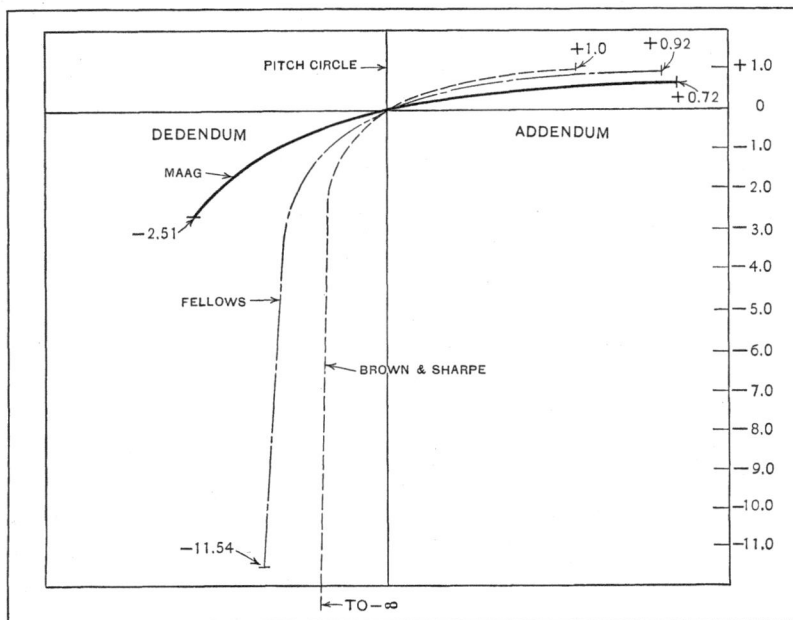

Fig. 19.   Sliding Diagrams of 12/12 Tooth Pinions Shown in Fig. 18.

For the second example, we will take a twelve-tooth pinion running with a twenty-four tooth gear.

These have the following values:

	B. & S.	Fellows	Maag
Number of teeth	12	12	12
Pitch radius	6.000	6.000	6.000
Pressure angle	14° 30′	20°	23° 14′
Radius of addendum circle	7.000	6.800	7.030
Radius of dedendum circle	5.000	5.200	5.343
Clearance	.157	.200	.140
Working depth of tooth	2.000	1.600	1.687
Radius of base circle	5.809	5.638	5.513
Circular pitch	3.1416	3.1416	3.1416
Number of teeth	24	24	24
Pitch radius	12.000	12.000	12.000
Radius of addendum circle	13.000	12.800	12.657
Radius of dedendum circle	11.000	11.200	10.970
Radius of base circle	11.618	11.276	11.027

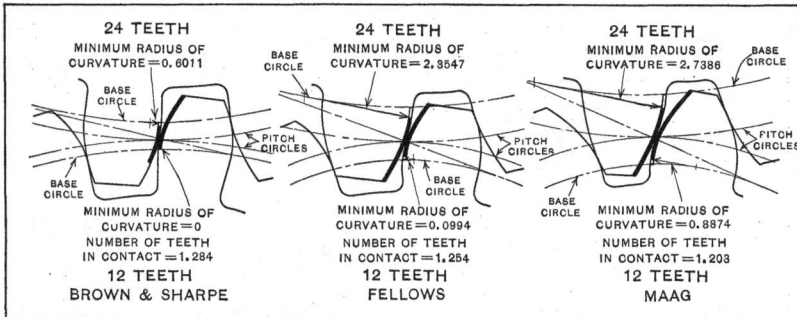

Fig. 20. Examples of 12/24 Tooth Gears.

Fig. 20 shows these gears. The same general differences in the characteristics of the three standards are apparent as before. The sliding diagrams are shown in Fig. 21. As the sliding conditions are different on the gear and pinion, two charts are shown. Here again, the curves for the Maag gears are flatter than the others.

For the last example, we will take a sixteen-tooth pinion running with a twenty-four-tooth gear. These have the following values:

	B. & S.	Fellows	Maag
Number of teeth	16	16	16
Pitch radius	8.000	8.000	8.000
Pressure angle	14° 30′	20°	21° 55′
Radius of addendum circle	9.000	8.800	9.000
Radius of dedendum circle	7.000	7.200	7.094
Clearance	.157	.200	.147
Working depth of tooth	2.000	1.600	1.759
Radius of base circle	7.745	7.518	7.423
Circular pitch	3.1416	3.1416	3.1416
Number of teeth	24	24	24
Pitch radius	12.000	12.000	12.000
Radius of addendum circle	13.000	12.800	12.760
Radius of dedendum circle	11.000	11.200	10.854
Radius of base circle	11.618	11.276	11.133

These gears are shown in Fig. 22, their sliding diagrams in Fig. 23. Here again the sliding curves for the Maag gears are much flatter than the others.

39

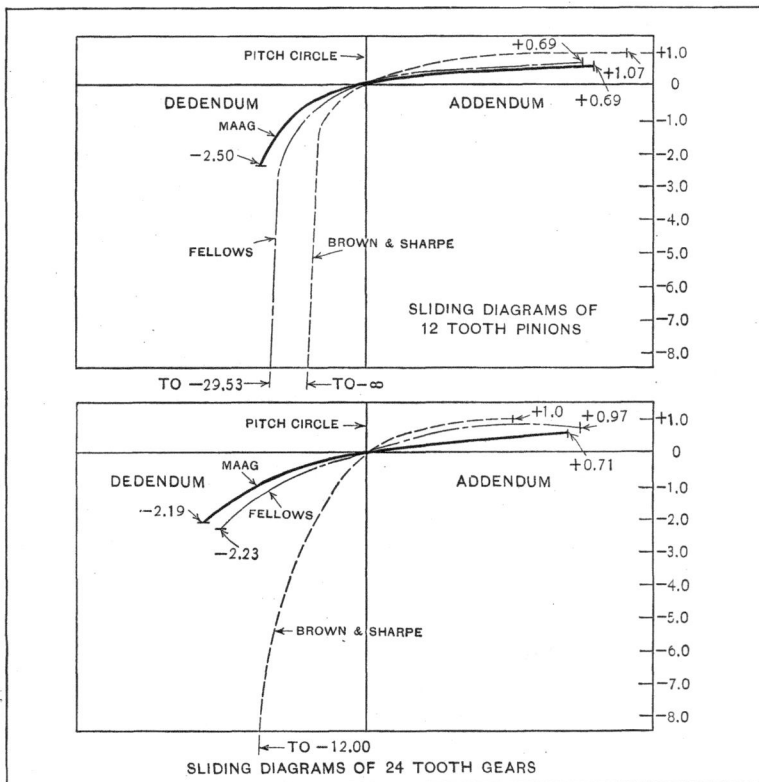

Fig. 21.   Sliding Diagrams of 12/24 Tooth Gears Shown in Fig. 20.

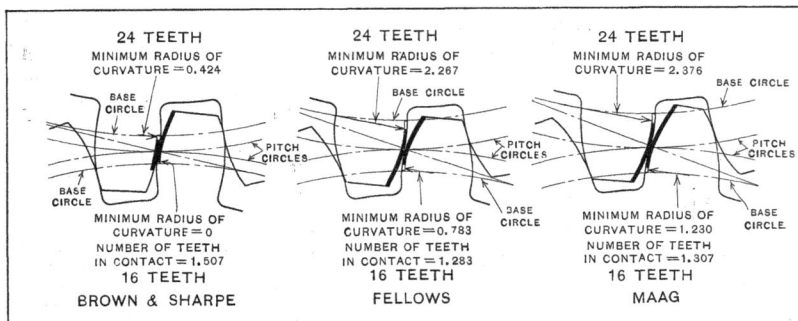

Fig. 22.   Examples of 16/24 Tooth Gears.

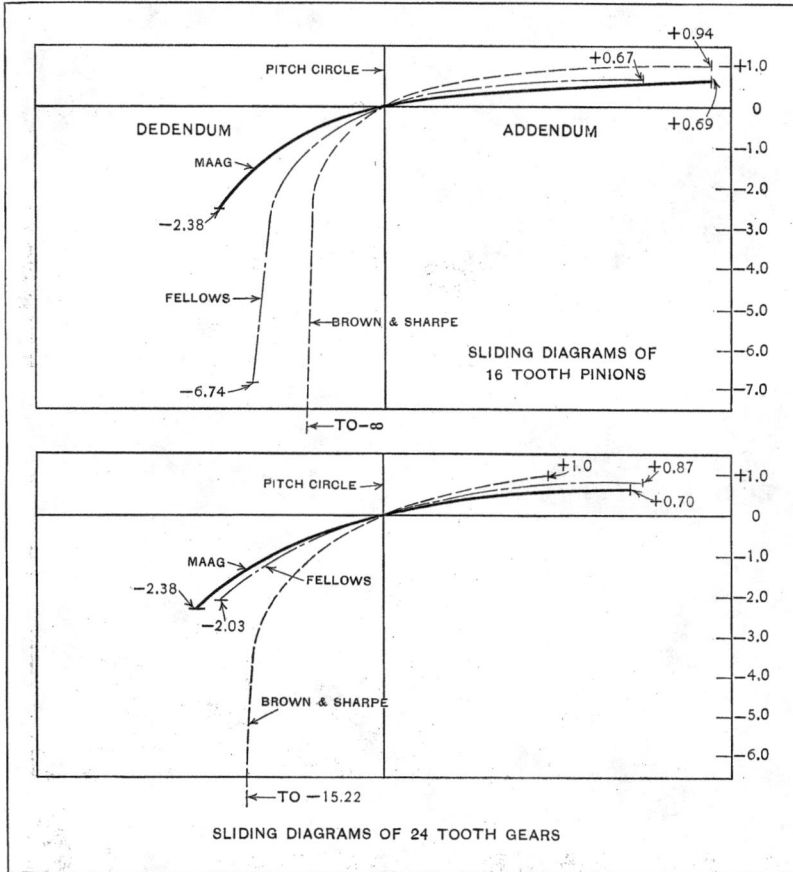

Fig. 23.   Sliding Diagrams of 16/24 Tooth Gears Shown in Fig. 22.

These few examples showing small numbers of teeth have been purposely chosen because the effect of improper tooth design is more pronounced than when larger numbers of teeth are involved.

Maag Gear Shaping Machine.

CHAPTER III

Methods of Production

Form Milling

THE production of gears by form milling is extensively employed because in this way they can be made without any special equipment other than the form cutters required. For general jobbing and repair work, a standard milling machine equipped with a dividing head is all that is needed.

This method also lends itself readily to production of gears in quantity. Oftentimes a manufacturing milling machine is equipped with a special automatic indexing fixture for this work.

Standard gear cutting machines are also on the market for milling gears. These are made in many types. Some have single work arbors while others have several and thus cut several gears or stacks of gears at once. All these machines are automatic. Fig. 24 shows a single arbor gear milling machine.

The milling cutters are form cutters relieved so that they can be sharpened by grinding the face of the tooth without losing their form. Such a cutter is shown in Fig. 25.

Theoretically, a different cutter is required for each gear of a different number of teeth. A series of cutters has been developed, however, for the 14½-degree standard tooth form so that only eight cutters are required for each pitch. These are adapted to cut from a pinion of twelve teeth to a rack, and are numbered respectively 1, 2, 3, 4, etc.

Fig. 24.   Brown & Sharpe Form Milling Gear Cutting Machine.

No. 1 will cut gears from 135 teeth to a rack
No. 2 will cut gears from 55 teeth to 134 teeth
No. 3 will cut gears from 35 teeth to 54 teeth
No. 4 will cut gears from 26 teeth to 34 teeth
No. 5 will cut gears from 21 teeth to 25 teeth
No. 6 will cut gears from 17 teeth to 20 teeth
No. 7 will cut gears from 14 teeth to 16 teeth
No. 8 will cut gears from 12 teeth to 13 teeth

These cutters are usually made of correct form for gears of the smallest number of teeth in their range.   Thus a No. 5 cutter is correct for 21 teeth, and approximates very closely the form for the others in the range.   These cutters are satisfactory where the conditions are not severe.   For more exacting work, cutters of the half numbers listed above are furnished as follows:

No. 1½ will cut gears from 80 teeth to 134 teeth
No. 2½ will cut gears from 42 teeth to 54 teeth
No. 3½ will cut gears from 30 teeth to 34 teeth
No. 4½ will cut gears from 23 teeth to 25 teeth
No. 5½ will cut gears from 19 teeth to 20 teeth
No. 6½ will cut gears from 15 teeth to 16 teeth
No. 7½ will cut gears of 13 teeth

Cutters are made of correct form for any tooth numbers and any pressure angle when required. This is an expense that is often justified for the production of a large quantity of gears.

Fig. 25. Form Cutter for Gears.

Great care must be exercised in setting up a milling machine to cut gears if reasonably accurate results are desired. The center of the cutter must be accurately aligned with the center of the blank. For this reason, the form cutters often have a line graved on them to locate the center of the cutter profile. The cutter must also be set accurately to the correct depth. Any mislocation from the correct relative positions of the cutter and gear blanks will spoil the gear.

Form milling of gears is extensively employed in the production of many of the ordinary gears which meet only the ordinary conditions. The process is also used to rough out gear blanks which are later finished by some generating process, and for miscellaneous jobbing and repair work. Gears cut by form milling are seldom, if ever, run at pitch line velocities exceeding 2,000 feet a minute.

Hobbing and Hobs

This process consists of revolving and advancing a worm shaped cutter through a revolving blank. The ratio between the speed of the hob and the blank is determined by the number of threads on the hob and the number of teeth required in the gear.   If a single threaded hob is used to cut a twenty-four tooth gear, the hob revolves twenty-four times while the gear blank revolves once.   If a double threaded hob is used, the hob revolves twelve times while the gear blank revolves once.

In general, the hob is set at such an angle in relation to the teeth of the gear that the helix at the middle of the tooth on the hob is tangent to the side of the gear teeth.   Sometimes, however, hobs are so made that their axis is set square with the axis of the gear.   A typical hobbing machine is shown in Fig. 26.

Hobbing is a generating process.   It has the advantage of requiring only one hob to cut gears of the same pitch with any number of teeth.   For quantity production in particular,

Fig. 26.   Barber-Colman Hobbing Machine.

hobbing is one of the most rapid methods. All motions are continuous, and, except for the feeding of the hob through the blanks, all motions are rotary. Such continuous rotary motions are ideal as far as mere rapid operation is concerned.

The usual hob consists of a straight-sided worm-shaped cutter, its axial section representing the developed form of the generating rack. If the axis of the hob is set square with the axis of the gear blank, its axial section represents the generating rack. As the hob is usually set at an angle to the blank, the angle of the sides of the hob teeth are altered accordingly.

Thus, let,

α = angle of side of generating rack;

β = angle at which hob is set;

α'' = angle of side of hob tooth.

Whence,

$$\text{Tan } \alpha'' = \frac{\tan \alpha}{\cos \beta}.$$

Let

P = circular pitch of rack;

K = lead of hob;

$$K = \frac{P}{\cos \beta} \text{ for single-threaded hob;}$$

$$K = \frac{2P}{\cos \beta} \text{ for double-threaded hob;}$$

etc.

Correct Profile of Hobs

An involute gear is generated from a straight-sided rack. Theoretically, a hob which will generate a true involute gear will also generate a straight-sided rack. Although, for practical reasons, a hob would never be used to cut a rack, we will use one in the study of the hob, because its straight line profile is the easiest example to consider.

The simplest example of such a hob is a spiral gear of one tooth (when the hob has but a single thread) running with a straight-sided rack. For the present, all consideration of backing off, nature of cutting flutes, etc., will be ignored. This hob consists of an infinite number of spur gears twisted uniformly to give the lead of the spiral.

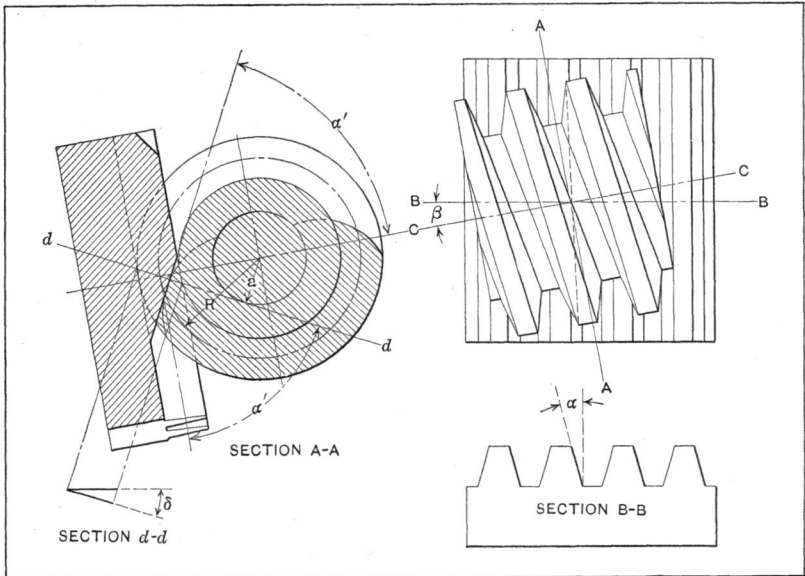

Fig. 27.   Rack and Hob.

Such a hob and rack are illustrated in Fig. 27.   In this illustration let

a = radius of base circle of involute in section $A-A$;

R = radius of pitch circle of involute;

a = angle of side of straight rack tooth;

a' = pressure angle of involute in section $A-A$, which is also the projected angle of a in section $A-A$;

β = angle at which hob is set.

In this illustration, β is also the helix angle of the hob at the radius R.

M = module of rack;

P = circular pitch of rack $= \pi M$;

K = lead of hob.

Referring to Fig. 27, we have

$$\tan a' = \frac{\tan a}{\sin \beta} ;$$

$$K = \frac{P}{\cos \beta} ;$$

$$\tan \beta = \frac{K}{2\pi R} = \frac{P}{2\pi R \cos \beta} = \frac{M}{2R \cos \beta} .$$

Whence

$$\sin \beta = \frac{M}{2R} ,$$

or

$$R = \frac{M}{2 \sin \beta}$$

Referring to section $A-A$ in Fig. 27, the line $d-d$ is perpendicular to the side of the developed section of the rack tooth and passes through the intersection of the centerline and pitch circle of the hob. This line $d-d$ is therefore the line of action of the involute and rack. We have already seen that contact between an involute and a straight line takes place only along this line of action. The circle to which this line $d-d$ is tangent will be the base circle of the involute. Whence,

$$a = R \cos \alpha'.$$

Considering now the infinite number of successive sections of the involute which are twisted uniformly to make the hob, we see that contact between the straight-sided rack and hob must be in a straight line, because the projection of the line of contact in the plane $A-A$ is the straight line $d-d$, and this line of contact lies in the plane of the straight-sided rack tooth. If the angle of this line of contact with a plane perpendicular to the axis of the hob is determined, we generate a hob of the required form by revolving this line of contact, keeping it tangent to a cylinder of the same diameter as the base circle of the involute in section $A-A$ and advancing it uniformly according to the lead of the hob.

Referring to Fig. 27, in section $A-A$ this line is at the angle α' from the horizontal. In the elevation $B-B$ it must be at the angle α from the vertical because this line of contact must be in the plane of the side of the rack. The angle δ in section $d-d$ is the desired angle.

Fig. 28 shows only the lines required to establish this angle. It is merely a problem in Descriptive Geometry. The contact line $d-d$ is shown in heavy lines in four projections. The projection lines enable their construction to be followed. The values of the lengths of the different lines are indicated in the figure. From these we have

$$\tan \delta = \frac{(\tan^2 \alpha + \sin^2 \beta) \sin \alpha'}{\cos \beta \tan \alpha}.$$

But

$$\sin \alpha' = \frac{a \tan \alpha}{R \sin \beta}.$$

Substituting this in the above equation, we have

$$\tan \delta = \frac{a (\tan^2 \alpha + \sin^2 \beta)}{R \sin \beta \cos \beta}.$$

49

But
$$\tan^2 \alpha = \sin^2 \beta \tan^2 \alpha'.$$
Substituting, we have

$$\tan \delta = \frac{a \sin \beta \, (\tan^2 \alpha' + 1)}{R \cos \beta} = \frac{a \tan \beta \, (\tan^2 \alpha' + 1)}{R}.$$

$$\tan \beta = \frac{K}{2\pi R}.$$

Substituting, we have,

$$\tan \delta = \frac{a K \, (\tan^2 \alpha' + 1)}{2\pi R^2}.$$

$$\tan^2 \alpha' + 1 = \frac{\sin^2 \alpha' + \cos^2 \alpha'}{\cos^2 \alpha'} = \frac{1}{\cos^2 \alpha'}$$

Substituting, we have

$$\tan \delta = \frac{a K}{2\pi R^2 \cos^2 \alpha'}$$

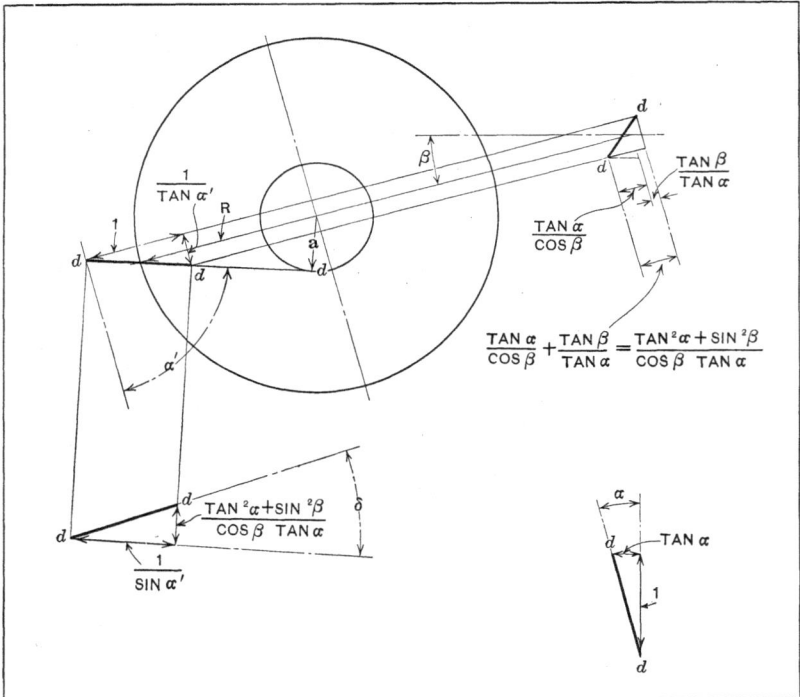

Fig. 28.   Diagram of Contact Between Rack and Hob.

But
$$R \cos \alpha' = \mathrm{a};$$
whence
$$\tan \delta = \frac{K}{2\pi \mathrm{a}}.$$

But the tangent of the helix angle of the hob at radius "a" is equal to $\frac{K}{2\pi \mathrm{a}}$. Therefore the surface of a hob which will generate a straight-sided rack or an involute gear is generated by revolving and advancing with a uniform lead, a straight line inclined at any given angle to the plane perpendicular to the axis, this line remaining tangent to a cylinder of such a diameter that the helix angle at that diameter is the same as the angle between the generating line and a plane perpendicular to the axis. Such a surface forms an involute helicoid. This feature of involute helicoidal surfaces makes it simple to calculate not only hobs but also helical and spiral gears whose surfaces are involute helicoids.

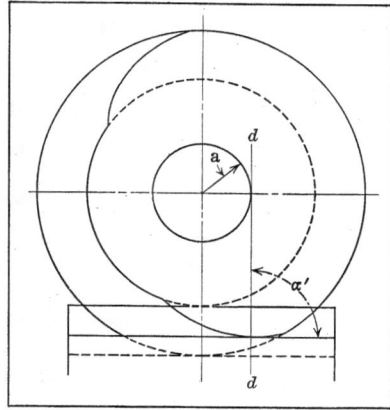

Fig. 29. Diagram of Hob Set Square with Rack.

The angle β at which the hob is set may be chosen, within certain practical limits, at random and a hob can be developed to cut a straight-sided rack, or involute gear, of the desired pitch. The angle β may be reduced to zero which would result in the conditions shown in Fig. 29.

Referring to previous equations,
$$\tan \alpha' = \frac{\tan \alpha}{\sin \beta}.$$
$$\beta = 0.$$
Whence
$$\sin \beta = 0;$$
$$\tan \alpha' = \frac{\tan \alpha}{0} = \infty.$$
Whence
$$\alpha' = 90°;$$
$$R = \frac{M}{2 \sin \beta} = \frac{M}{0} = \infty;$$

$$\tan \delta = \frac{(\tan^2 \alpha + \sin^2 \beta) \sin \alpha'}{\cos \beta \tan \alpha}$$

$\beta = 0$, whence $\sin \beta = 0$, and $\cos \beta = 1$
$\alpha' = 90°$, whence $\sin \alpha' = 1$

$$\tan \delta = \frac{(\tan^2 \alpha + 0) \times 1}{1 \times \tan c} = \tan \alpha.$$

Whence

$$\delta = \alpha.$$

Thus, when the angle β at which the hob is set is equal to zero, the angle δ of the generating line is equal to the angle of the rack. The pressure angle in section $A–A$ will be 90 degrees and the radius of the pitch circle will be infinity.

When the hob is set with a helix tangent to the tooth of the rack or gear at any given diameter of the hob, the thickness of the hob tooth at that point is equal to the width of the space at the mating portion of the rack or gear. In all other cases the space cut by the hob tooth will be greater than the thickness of the hob tooth. Therefore, if a pre-determined width of space must be maintained, the thickness of the hob tooth must be reduced accordingly. This reduction in the thickness of the hob tooth is the principal limitation in regard to selecting the angle at which to set the hob.

Before proceeding further, it is of interest to determine the effect on the form of an involute gear tooth of a variation in the setting angle β.

Referring to previous equations we have the following:

$$K = \frac{P}{\cos \beta};$$

whence

$$P = K \cos \beta;$$

$$\tan \delta = \frac{K}{2\pi a};$$

$$a = R \cos \alpha'$$

$$\tan \alpha' = \frac{\tan \alpha}{\sin \beta};$$

$$M = \frac{P}{\pi};$$

$$R = \frac{M}{2 \sin \beta} = \frac{P}{2\pi \sin \beta};$$

whence

$$\frac{\sin a'}{\cos a'} = \frac{\tan a}{\sin \beta};$$

$$\frac{1 - \cos^2 a'}{\cos^2 a'} = \frac{\tan^2 a}{\sin^2 \beta};$$

whence

$$\cos a' = \frac{\sin a}{\sqrt{\tan^2 a + \sin^2 \beta}};$$

$$a = R \cos a' = \frac{P}{2\pi \sqrt{\tan^2 a + \sin^2 \beta}};$$

$$\tan \delta = \frac{K}{2\pi a} = \frac{K \sqrt{\tan^2 a + \sin^2 \beta}}{P} = \frac{\sqrt{\tan^2 a + \sin^2 \beta}}{\cos \beta};$$

$$\tan^2 \delta = \frac{\tan^2 a + \sin^2 \beta}{\cos^2 \beta} = \frac{\tan^2 a + 1 - \cos^2 \beta}{\cos^2 \beta};$$

$$\cos^2 \beta (\tan^2\delta + 1) = \tan^2 a + 1;$$

$$\frac{\cos^2 \beta}{\cos^2 \delta} = \frac{1}{\cos^2 a};$$

$$\cos^2 \delta = \cos^2 a \cos^2 \beta.$$

whence

$$\cos \delta = \cos a \cos \beta.$$

It will be seen from the foregoing equation that δ is dependent only on the angles a and β.

With an involute spur gear, if the pitch diameter is changed while the involute profile remains unchanged, the module and pressure angle change accordingly and maintain the following relationship:

Let $M_1 =$ original module of gear;

$M_2 =$ changed module of gear;

$a_1 =$ original pressure angle;

$a_2 =$ changed pressure angle;

$M_1 \cos a_1 = M_2 \cos a_2.$

whence

$$M_1 = \frac{M_2 \cos a_2}{\cos a_1};$$

If the change in the module is caused by a change in the angular setting of the hob β, the new module will have the following relationship with the old:

Let $\beta_1 =$ original setting of hob;

$\beta_2 =$ changed setting of hob;

53

$$K = \frac{P}{\cos \beta} = \frac{\pi M_1}{\cos \beta_1} = \frac{\pi M_2}{\cos \beta_2};$$

whence

$$\frac{\pi M_1}{\cos \beta_1} = \frac{\pi M_2}{\cos \beta_2};$$

$$\frac{M_1}{\cos \beta_1} = \frac{M_2}{\cos \beta_2};$$

$$M_1 = \frac{M_2 \cos \beta_1}{\cos \beta_2};$$

whence

$$\frac{M_2 \cos \alpha_2}{\cos \alpha_1} = \frac{M_2 \cos \beta_1}{\cos \beta_2};$$

$$\cos \alpha_1 \cos \beta_1 = \cos \alpha_2 \cos \beta_2.$$

Referring to previous equation:

$$\cos \delta = \cos \alpha \cos \beta = \cos \alpha_1 \cos \beta_1 = \cos \alpha_2 \cos \beta_2.$$

Therefore, regardless of the angle at which a hob is set, it produces the same involute on the gear being cut. The thickness of the space and tooth will change as the setting angle is altered, but this will be the only difference.

It is evident that this freedom in choosing the angle at which the hob may be set is of great practical value. Machine settings are never mathematically exact, and do not need to be for hobs. The hob may be set to depth and angle by trial to secure the desired tooth thickness, and no theoretical error will develop. Furthermore, the thickness of the hob tooth itself is not important as long as it is not too great. A slight change in its angular setting on the machine will automatically compensate for any errors in thickness.

Shaping With Pinion-Shaped Cutters.

Another way to generate involute gears is to use a pinion-shaped cutter in a special shaping or planing machine, as shown in Fig. 30. Both the cutter and gear blank are revolved the proper distance between each stroke of the machine. This method is extensively used in the automotive industries for finishing gears which have been roughed out on a gear milling or hobbing machine.

Fig. 30. Fellow's Gear Shaping Machine.

Here, as in hobbing, the cutter is a vital factor in getting accurate results.   Its diameter is limited.   These cutters are made either three or four inches in pitch diameter, most extensively three. See Fig. 31.

Fig. 31.   Pinion-shaped Cutter.

As the cutter is usually smaller than the mate of the gear being cut, these gears need more clearance than those produced by other methods. See Fig. 32.   The corner of the tooth of the cutter develops a fillet in the generated gear which sometimes extend into the active profile.

This method has the same advantage as hobbing—namely that one cutter cuts gears with any number of teeth. The small diameter of the cutter, however, introduces a serious handicap.

55

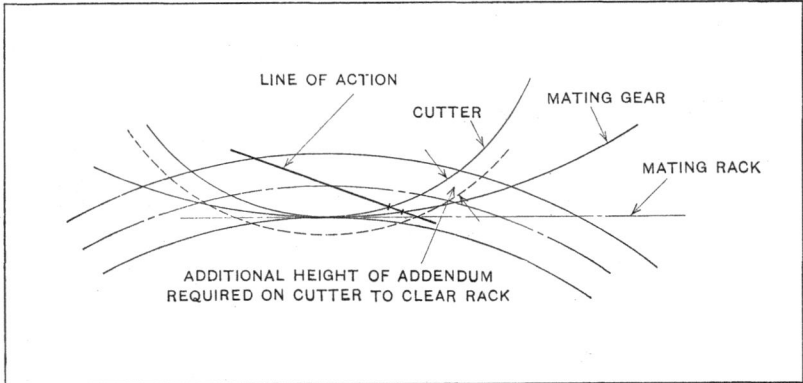

Fig. 32.   Diagram Showing Clearance Required When Using Pinion-shaped
Cutter.

On coarse pitches, the number of teeth in the cutter is so small that its teeth are made with radial flanks. This modifies the involute form on the generated gears. Sometimes this modification is still further accentuated by having these straight flanks not quite radial.

It is possible to overcome this disadvantage, but to do so requires several cutters to cut gears with different numbers of teeth.

As an example, we will consider a three inch pitch diameter cutter to cut a $\frac{4}{5}$ pitch Fellow's Stub Tooth Gear. This cutter would have the following dimensions:

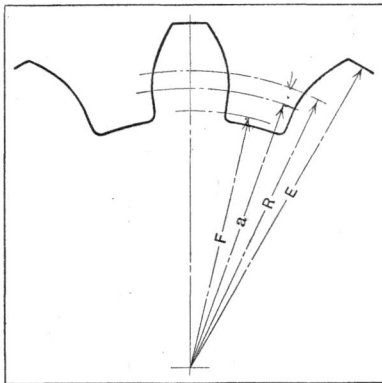

Fig. 33.   Pinion-shaped Cutter of
Usual Design.

Let E = outside radius;
R = radius of pitch circle;
F = radius of dedendum circle;
a = radius of base circle;
a = pressure angle;
$R = 1.500$;
$E = R + .250 = 1.750$.
$F = R - .200 = 1.300$;
a $= R \cos a = 1.4095$;
$P = .7854$;
$a = 20°$.

The form of the tooth of this cutter is shown in Fig. 33. The radial flanks extend over about one-quarter of the tooth profile.

To eliminate these radial flanks, the dedendum circle of the cutter is moved up to or beyond the base circle. This has been done in Fig. 34, which shows a three-inch pitch diameter cutter with the following values:

$R = 1.500$;
$E = 1.880$;
$F = 1.430$;
$a = R \cos \alpha = 1.4095$;
$P = .7854$;
$\alpha = 20°$.

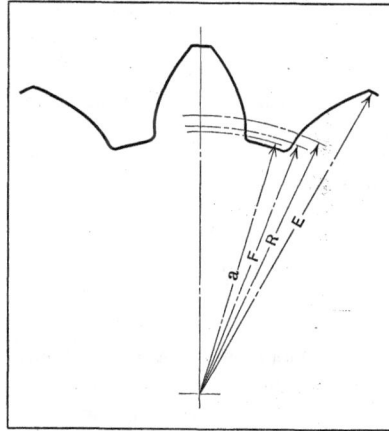

Fig. 34. Pinion-shaped Cutter of Improved Design.

To cut these gears with different numbers of teeth and keep the thickness of tooth and space on the pitch line equal, the thickness of the cutter tooth on the pitch line must vary. In Fig. 35 this cutter generates a gear to run at a 20° pressure angle with 18 teeth. The dimensions of this gear when running with a pressure angle of 20° are as follows:

$R = 2.250$;
$E = 2.450$;
$F = 2.050$;
$f = .050$
$P = .7854$;
$\alpha = 20°$.
$a = R \cos \alpha = 2.1143$.

The center distance between the standard cutter with radial flanks and this gear is 3.750. The center distance between the cutter shown in Fig. 34 and this gear, however, is $2.450 + 1.430 = 3.880$. Thus in Fig. 35, the pressure angle between the cutter of full involute form and the 18-tooth gear is determined as follows:

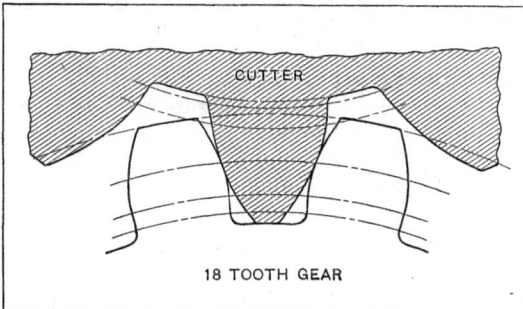

Fig. 35. Improved Pinion-shaped Cutter with Standard Gear.

a_1 = radius of base circle of cutter;
a_2 = radius of base circle of gear;
C = center distance;
a' = pressure angle between cutter and gear;

$$\cos a' = \frac{a_1 + a_2}{C}.$$

Whence
$\cos a' = .90821;$
$a' = 24°\text{-}44'\text{-}26''.$

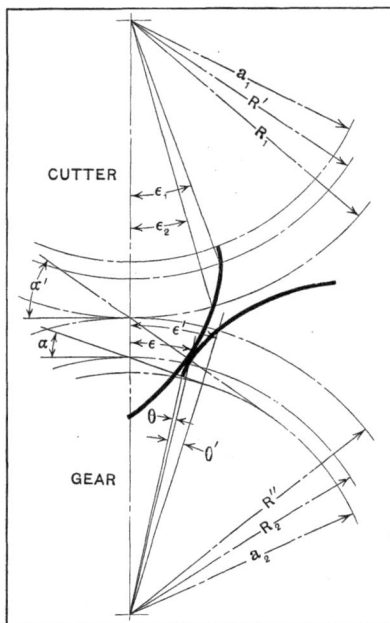

CUTTER

GEAR

We must next determine the proper thickness of the tooth of the cutter at the pitch line. To do this we use the equations of the involute curve previously given. In Fig. 36, let

θ = angle of radius vector for pressure angle of a;

θ' = angle of radius vector for pressure angle of a';

ϵ = arc of half the tooth space on a pitch line of gear;

Fig. 36. Diagram for Developing Formulae.

ϵ' = arc of half the tooth space on a' pitch line of gear;

ϵ_1 = arc of half the tooth thickness on a pitch line of cutter;

ϵ_2 = arc of half the tooth thickness on a' pitch line of cutter;

R_1 = radius of pitch circle of a pressure angle on cutter;

R' = radius of pitch circle of a' pressure angle on cutter;

R_2 = radius of pitch circle of a pressure angle on gear;

R'' = radius of pitch circle of a' pressure angle on gear;

a_1 = radius of base circle of cutter;

a_2 = radius of base circle of gear;

P = circular pitch of gear or cutter with pressure angle of a

Whence

$$\epsilon = \frac{P}{4R_2};$$

$$\theta = \tan a - a;$$
$$\theta' = \tan a' - a';$$
$$\epsilon' = \epsilon + (\theta' - \theta);$$
$$R''\epsilon' = R'\epsilon_2;$$

$$\epsilon_2 = \frac{R''}{R'}\epsilon'.$$

But

$$\frac{R''}{R'} = \frac{a_2 \cos a'}{a_1 \cos a'} = \frac{R_2}{R_1}.$$

Whence

$$\epsilon_1 = \epsilon_2 + (\theta' - \theta) = \frac{R_2}{R_1}[\epsilon + (\theta' - \theta)] + (\theta' - \theta);$$

$$\epsilon_1 = \frac{R_2}{R_1}\epsilon + \frac{R_2}{R_1}(\theta' - \theta) + (\theta' - \theta);$$

$$\epsilon_1 = \frac{R_2}{R_1}\epsilon + (\theta' - \theta)\left(\frac{R_2}{R_1} + 1\right).$$

Thickness of tooth of cutter on pitch line $a = 2R_1\epsilon_1$.

Referring again to Fig. 35, we have

$R_1 = 1.500$;

$R_2 = 2.250$;

$P = .7854$;

$a = 20°$;

$a' = 24°\text{-}44'\text{-}26''$;

$\theta = \tan a - a = .01490438$;

$\theta' = \tan a' - a' = .02900232$;

$\theta' - \theta = .01409794$;

$$\epsilon = \frac{P}{4R_2} = .08726646;$$

$$\epsilon_1 = \frac{R_2}{R_1}\epsilon + (\theta' - \theta)\left(\frac{R_2}{R_1} + 1\right) = .16614454.$$

Thickness of cutter tooth at $R_1 = 2R_1\epsilon_1 = .4984$.

The proper thickness of the cutter tooth at this point to cut gears with different numbers of teeth is found in similar manner. Thus we obtain the following:

No. of Teeth	Thickness of Cutter Tooth at 3″ Diameter	No. of Teeth	Thickness of Cutter Tooth at 3″ Diameter
12	.5009	22	.4973
13	.5004	23	.4970
14	.5000	24	.4967
15	.4995	25	.4965
16	.4991	26	.4962
17	.4988	27	.4960
18	.4984	28	.4958
19	.4981	29	.4956
20	.4978	30	.4955
21	.4975		

It will be seen from the above tabulation that the variation in the thickness of the cutter tooth is slight, and may be taken care of in one of two ways. By the first method the cutter remains constant, and the depth of the teeth in the different gears must vary. Thus if the cutter is correct for a thirty-tooth gear, the depth of tooth in a twelve-tooth gear is about .008″ greater than the standard depth. By the second method, a series of cutters can be made so that the depth is kept within about .002″ of standard. Thus one cutter covers a range of from twelve to sixteen teeth, a second cutter can be made to cover a range of from seventeen to twenty-two teeth, a third to cover from twenty-three to thirty teeth, etc.

The sides of the cutter must be relieved to give a proper cutting edge, but must retain the correct form at the proper diameter. This is accomplished by making these relieved surfaces involute helicoids the relief on one side of the tooth being right hand, while the relief on the other is left hand. As a matter of fact, these cutters are nothing more than multiple-threaded hobs with an infinite lead, the cutting action being obtained by reciprocating the cutter instead of revolving it. The tops of the teeth are also relieved, in such a way that the width at the top of the cutter tooth remains approximately the same. As the face of the cutter is ground back when re-sharpened, a different portion of the involute profile of the cutter comes into action. This, however, as we have seen before, in considering the action of one involute with another, does not change the form of the gear produced.

The intersection of the side relieved surfaces of the cutter with a plane perpendicular to its axis is an involute curve. The polar equation of this curve, as shown before, is as follows:

$$\theta = \sqrt{\left(\frac{r}{a}\right)^2 - 1} - \arctan \sqrt{\left(\frac{r}{a}\right)^2 - 1}.$$

The front face of the pinion cutter, however, is not a plane but a conical surface.

Fig. 37 shows a pinion cutter in front elevation and axial section respectively. Line A–B is the intersection line of the conical front surface with the drawing plane. The angle of this line against a plane which is perpendicular to the axis of the pinion cutter has been designated γ. The axial elevation of any point of the conical surface from the apex A is therefore

$$z = r \tan \gamma.$$

Involute intersection curves are obtained by intersecting the relieved side surfaces of the pinion cutter with parallel planes which are perpendicular to the axis of the cutter. If these planes are equally spaced, the involute intersection curves will be twisted uni-

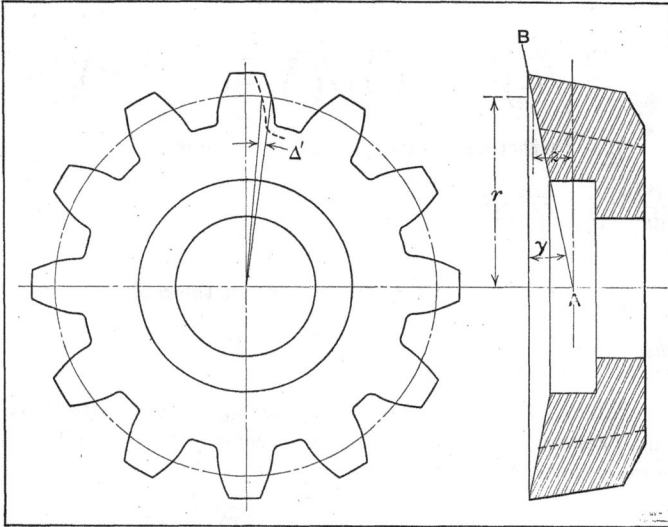

Fig. 37.   Pinion-shaped Cutter and Relief.

formly from some common centerline, the amount of this twisting depending upon the lead K of the involute helicoidal relieved surfaces.   The twisting or turning angle Δ' in Fig. 37 is dependent upon the axial distance of the intersecting plane from the apex A.   Thus, if this distance is equal to the lead of the involute helicoidal relieved surfaces, the turning angle $\Delta' = 2\,\pi$.

In a fixed polar co-ordinate system with the origin at A, the equation of the involute which is intersected at the distance z from A becomes:

$$\theta = -\,2\pi\frac{z}{K} + \sqrt{\left(\frac{r}{a}\right)^2 - 1} - \arctan\sqrt{\left(\frac{r}{a}\right)^2 - 1}.$$

Introducing the relation:

$$z = r\tan\gamma$$

we get the equation of the intersection curve.

$$\theta = -\,\frac{2\pi}{K}\tan\gamma\,r + \sqrt{\left(\frac{r}{a}\right)^2 - 1} - \arctan\sqrt{\left(\frac{r}{a}\right)^2 - 1}.$$

Let $\dfrac{2\pi}{K}\tan\gamma = G$, we have

$$\theta = -\,G\,r + \sqrt{\left(\frac{r}{a}\right)^2 - 1} - \arctan\sqrt{\left(\frac{r}{a}\right)^2 - 1}.$$

The derivations of this equation are as follows:

61

$$\frac{d\theta}{dr} = -G + \frac{\frac{r}{a^2}}{\sqrt{\left(\frac{r}{a}\right)^2 - 1}} - \frac{\frac{r}{a^2}}{\left(\frac{r}{a}\right)^2\sqrt{\left(\frac{r}{a}\right)^2 - 1}} = -G + \frac{1}{r}\sqrt{\left(\frac{r}{a}\right)^2 - 1}.$$

The further derivatives depend no more on G.

$$\frac{d^2\theta}{dr^2} = \frac{a}{r^3}\frac{1}{\sqrt{\left(\frac{r}{a}\right)^2 - 1}}.$$

The inclination of the tangent against the radius is

$$\tan \Psi = r\frac{d\theta}{dr}.$$

This inclination angle at the pitch line of the cutter is equal to the pressure angle of the cutting profile. Thus, we have, when a = pressure angle of the cutting profile at pitch line:

$$\tan a = -Gr + \sqrt{\left(\frac{r}{a}\right)^2 - 1}.$$

If this inclination is to be correct at the pitch radius of the cutter, we have

a = pressure angle of cutting edges;
a' = pressure angle of involute helicoidal relieved surfaces;
a_1 = base circle of involute with pressure angle of a;
a_2 = base circle of involute with pressure angle of a';
R = radius of pitch circle;

$$\tan a = -GR + \sqrt{\left(\frac{R}{a_2}\right)^2 - 1} = \sqrt{\left(\frac{R}{a_1}\right)^2 - 1};$$

$$\tan a' = \sqrt{\left(\frac{R}{a_2}\right)^2 - 1} = \tan a + GR.$$

But

$$GR = \frac{2\pi R}{K}\tan \gamma;$$

$\dfrac{2\pi R}{K}$ = tangent of the angle of the helix of the involute helocoid at the pitch radius against the axis.

Let this angle be Σ, we have
$GR = \tan \gamma \tan \Sigma$;
$\tan a' = \tan a + \tan \gamma \tan \Sigma.$

We will take as an example a three-inch diameter cutter which must cut standard 20 degree stub tooth gears. We will assume that the angle of the relief at the top of the cutter is 7 degrees, and the angle of the conical front face is 5 degrees. This gives the following:

$R = 1.500;$

$\gamma = 5°;$

$a = 20°.$

If we assume that the angle of relief, Σ, at the pitch line of the cutter is equivalent to the angle of the top relief, we have

$$\tan \Sigma = \tan 7° \tan a.$$

Whence we have

$$\tan\ a' = \tan\ a + \tan \gamma \tan\ \Sigma = \tan a\ (1 + \tan 7°\ \tan \gamma) =$$
$$\tan 20°\ (1 + \tan 7° \tan 5°) = \tan 20°\ (1 + .01074226) = .36788011.$$
$$a' = 20°-11'-51''.$$

The profile of the cutting edge of a pinion cutter with a conical front face is not a true involute curve. We will examine it to determine the nature and amount of this error in form. The correct cutting edge has the equation

$$\theta = \sqrt{\left(\frac{r}{a_1}\right)^2 - 1} - \arctan\sqrt{\left(\frac{r}{a_1}\right)^2 - 1}.$$

We calculate three points on this curve as follows:

$r =$ outside radius, E;

$r =$ pitch radius, R;

$r =$ radius of base circle, a_1.

In the case of a three-inch pitch diameter cutter of 4/5 pitch, $20°$ pressure angle, we have the following:

When $r = a_1,$

$$\theta = \sqrt{\left(\frac{a_1}{a_1}\right)^2 - 1} - \arctan\sqrt{\left(\frac{a_1}{a_1}\right)^2 - 1} = 0.$$

When $r = R = 1.5,$

$\theta = \tan 20° -$ arc $20° = .01490438.$

When $r = E = 1.75,$

$$a_1 = R \cos 20° = 1.4095389;$$

$$\left(\frac{r}{a_1}\right)^2 - 1 = \left(\frac{1.75}{1.4095}\right)^2 - 1 = .54142347;$$

$$\sqrt{\left(\frac{E}{a_1}\right)^2 - 1} = .735814.$$

Whence

$$\theta = .735814 - \arctan .735814 = .735814 - .634359$$
$$= .101455.$$

63

The actual cutting edge, however, has the equation

$$\theta = -Gr + \sqrt{\left(\frac{r}{a_2}\right)^2 - 1} - \arctan\sqrt{\left(\frac{r}{a_2}\right)^2 - 1}.$$

Thus, when

$$r = a_1,$$

$$\theta = -Ga_1 + \sqrt{\left(\frac{a_1}{a_2}\right)^2 - 1} - \arctan\sqrt{\left(\frac{a_1}{a_2}\right)^2 - 1}.$$

$$a_1 = 1.4095389;$$
$$a_2 = R \cos \alpha' = 1.4077621;$$

$$\left(\frac{a_1}{a_2}\right)^2 - 1 = .00252589;$$

$$\sqrt{\left(\frac{a_1}{a_2}\right)^2 - 1} = .0502583;$$

$$G = \frac{2\pi}{K} \tan \gamma;$$
$$\gamma = 5°;$$

$$\frac{2\pi R}{K} = \tan \Sigma = \tan 7° \tan 20°.$$

Whence

$$K = \frac{2\pi R}{\tan 7° \tan 20°} = 210.89268;$$

$$G = \frac{2\pi \tan \gamma}{210.89268} = 0026065745;$$

$$\theta = -Ga_1 + .0502583 - \arctan .0502583 = -.0035989$$

When

$$r = R_1,$$

$$\theta = -GR + \sqrt{\left(\frac{R}{a_2}\right)^2 - 1} - \arctan\sqrt{\left(\frac{R}{a_2}\right)^2 - 1}.$$

But

$$-GR + \sqrt{\left(\frac{R}{a_2}\right)^2 - 1} = \tan \alpha = .36397023;$$

$$\left(\frac{R}{a_2}\right)^2 - 1 = .13533489;$$

$$\sqrt{\left(\frac{R}{a_2}\right)^2 - 1} = .3678789;$$

$$\arctan .3678789 = .35251288;$$
$$\theta = .01145735$$

When $r = E$,

$$\theta = -GE + \sqrt{\left(\frac{E}{a_2}\right)^2 - 1} - \arctan \sqrt{\left(\frac{E}{a_2}\right)^2 - 1}$$

$$-GE = -.00456151;$$

$$\left(\frac{E}{a_2}\right)^2 - 1 = .54531694;$$

$$\sqrt{\left(\frac{E}{a_2}\right)^2 - 1} = .73845578;$$

$$\theta = .00456151 + .73845578 - \arctan .73845578$$
$$= .09782357.$$

The foregoing points on the two curves are tabulated below:

Value of r	θ For Involute Profile	θ For Actual Profile
a_1	0	—.0035989
R	+.01490438	+.01145735
E	+.101455	+.09782357

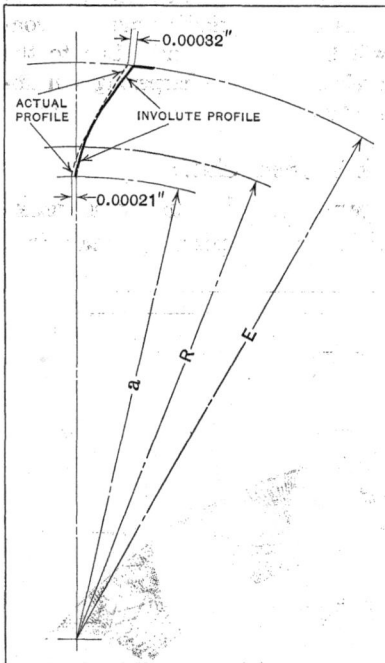

Fig. 38. Diagram of Error on Cutting Edge of Pinion-shaped Cutter.

To obtain the angular error of the actual profile, we must use the same polar co-ordinate system for both curves. Considering the point at R of both curves as being identical, we must move the points on the actual profile ahead the circular amount of .01490438—.01145735, which equals .00344703.

This gives the following tabulations:

Value of r	θ For Involute Profile
a_1	0
R	+.01490438
E	+.101455

θ For Actual Profile	Angular Error in Profile
—.0001519	—.0001519
+.01490438	0
+.01027060	—.000184

In order to measure the error as a length on the circumference of a circle, we must multiply the above angular errors by the radius of the corresponding circle. This gives the following errors for the profile of the actual cutting edge:

Value of r	Error in Cutting Profile in Inches
a_1 (1.4095389)	—.00021
R (1.500)	0
E (1.750)	—.00032

The actual cutting profile is plotted against a true involute profile in Fig. 38. The actual error is small, and is less than the other probable errors which develop in the mechanical processes of making and using these cutters. The actual cutting profile is less curved than the theoretically correct one. The cutter therefore removes somewhat more material than it ought to do. An error in this direction is preferable to one in the opposite direction which would leave too much metal on the gear teeth and develop interference.

Great care must be exercised when re-sharpening cutters of this type. A change in the angle of the conical front face alters the profile of the cutting edge so that it generates gears of slightly different pressure angle than originally. This conical grinding must be concentric, else the cutting edges will become eccentric. Due to the relieved surfaces of the cutter, carelessness in re-sharpening will impair the accuracy of the cutting edge.

Generating With Rack-Shaped Cutter

Involute gears can also be generated by using a rack-shaped cutter with a special shaping or planing machine.

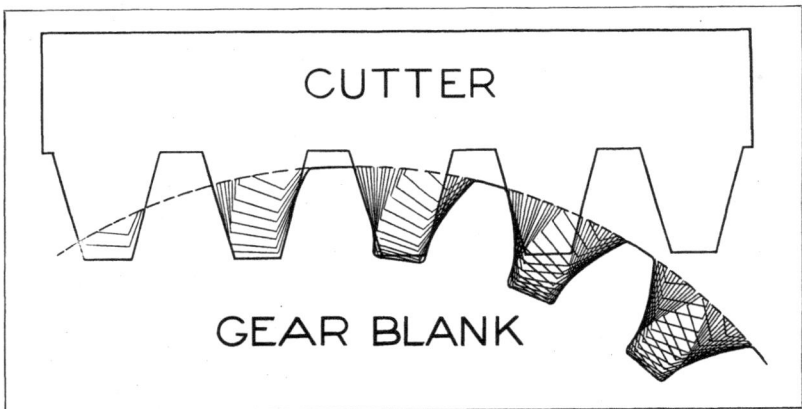

Fig. 39. Different Positions for Rack-shaped Cutter in Generating a Gear.

Between each stroke of the cutter the blank is revolved and advanced the proper distance along the face of the rack cutter. When the blank has advanced a distance equal to the circular pitch, it returns to its starting point without any rotary motion. Thus the successive teeth in the gear are generated by the same tooth in the rack cutter. Fig. 39 shows the different positions for a rack-shaped cutter in generating an involute spur gear. A machine cutting a large spur gear is shown on Page 42.

This method of generating gears has the same advantage as hobbing and cutting with pinion-shaped cutters in that one cutter cuts gears with any number of teeth. It is interesting to note that a pinion-shaped cutter is in effect a multiple-threaded hob with an infinite lead. The rack cutter is both a section of a hob of infinite diameter and a section of a pinion-shaped cutter of infinite diameter.

All the surfaces of a rack-shaped cutter are planes, so that the cutting edges are formed by the intersections of planes and are always straight lines. The proper angles of the rack cutters are readily computed, and no mathematical errors exist in the profiles of the cutting edges. Furthermore, the planes forming the several surfaces of the cutter can be machined and ground accurately and economically. This method of generating gears offers, therefore, the greatest possibilities of accuracy of any method. In all cases, the problem of the cutting tools is the crucial one. The machines used to generate the gears by any method can be made to practically equal accuracy.

Summary of Gear Generating Methods

The following is a summary of the various methods of generating involute spur gears:

Production

The hobbing method is the fastest. The continuous rotary motions employed in this process are ideal for speed. The pinion-shaped cutter method and the rack-shaped cutter method are virtually equal.

Flexibility of Cutting Tool

The hob is the most flexible of the three generating tools. To cut wider or narrower spaces in the gear, only its angular setting is altered. When generating teeth of special form with standard hobs, this characteristic is valuable. The rack-shaped cutter is next in order as regards flexibility.

Mathematical Accuracy

Mathematically the rack-shaped cutter is the best generating tool, the hob second. The mathematical profile of the hob can be generated as exactly as that of the rack-shaped tool, but when the problems of relief and cutting action are considered, the latter is much the better. The pinion-shaped cutter has a slight mathematical error in its cutting profile which cannot be corrected.

Ease of Making Accurate Tools

The rack-shaped cutter is the easiest one to make accurately because only plane surfaces are involved. The hob comes next in order. The pinion-shaped cutter is difficult to make accurately, because of the number of teeth which must have accurate profiles, be concentric and be uniformly spaced. When using a rack or a hob, all the teeth in the gear are generated by one tooth, but with a pinion-shaped cutter, the successive teeth in the gear are generated by the successive teeth in the cutter.

Cutting Action

The rack-shaped cutter offers better possibilities of getting the best cutting action, since the relief and rake can be chosen to suit conditions. The pinion-shaped cutter comes next, while the hob has the poorest cutting action. One side of the hob tooth can be made with an effective cutting edge, but the other side will have a negative rake which introduces bad cutting conditions.

It is apparent, then, that no one method has all the advantages. Specific requirements determine which is best to use.

CHAPTER IV

Methods of Testing Gears

Fig. 41. Brown & Sharpe Gear Tooth Vernier.

THE testing of gear tooth profiles, spacing, etc., is one of the most difficult inspection operations, and no one instrument is sufficient. Several devices are employed, each of which has been developed to measure particular features of the gear.

Gear Tooth Vernier

The gear tooth vernier, which is used to measure the thickness of the tooth, is shown in Fig. 41.

One objection to this instrument is that the actual contact between it and the product is a sharp corner, which soon wears away, giving a false reading. To overcome this, the instrument must be calibrated frequently, and the necessary correction applied to the reading.

These instruments cover a wide range of pitches, regardless of the pressure angle of the gears. The smallest one covers a range of from 20 to 2 diametral pitch, while a larger instrument covers the range of from 10 to 1 diametral pitch.

Fig. 42. Pratt & Whitney Gear Tooth Micrometer.

Gear Tooth Micrometer.

Another instrument used for the same purpose is shown in Fig. 42. The base of the gage represents a space of the basic rack of the gear tooth system, and a micrometer head measures the height of the tooth. The instrument makes contact with the gear on a flat face instead of on sharp corners. A separate instrument, however, is required for each pitch. They may be used on gears of various pressure angles, but extensive computations are involved. Practically, it is best to have a separate instrument for each pitch and each pressure angle. This instrument is primarily a gage for extensive production.

This gage can be made to cover a wider range, however,

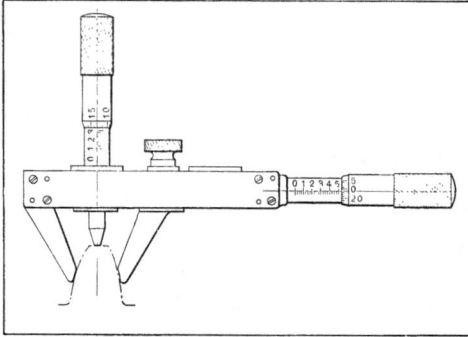

Fig. 43.   Adjustable Gear Tooth Micrometer.

simply by making the jaws adjustable as shown in Fig. 43.

Another way to measure the thickness of gear teeth is to place two ground pins or rolls of predetermined size between the teeth on opposite sides of the gears, and measure over these rolls with a micrometer, as shown in Fig. 44. The proper figure for this measurement is usually determined experimentally, as the calculations required are complex.

Testing Gears On Centers.

Gears are tested on centers for two purposes: First, to insure that they will run at the proper center distance with

Fig. 44.   Measuring Thickness of Teeth by Means of Rolls.

Fig. 45.   Morse Gear Testing Machine.

Fig. 46.   Saurer Gear Testing Machine.

suitable backlash; second, to test them for concentricity. Special fixtures or gages are often constructed to test the center distance between pairs of gears that are manufactured in large quantities. For general testing purposes, several machines of similar design are on the market. Such a machine is shown in Fig. 45. These testing stands comprise a bed with one fixed spindle and one movable spindle. A scale is mounted on the bed which may be read to one thousandth of an inch by means of a vernier plate, carried on the sliding head. The spindles are set a predetermined distance apart, and the gears are then mounted on the spindles. The amount of backlash may be measured by inserting thin feeler gages between the engaging teeth or by inserting pieces of paper until the gears are tight and then measure the thickness of the papers. If the gears are eccentric, a varying amount of backlash will be found at different positions of the gears.

Fig. 47.   Testing Concentricity of Gear with Pin and Indicator.

Another method of measuring the backlash is to adjust the spindles of the machine until the gears are tight together. The difference between this center distance and the correct center distance multiplied by twice the tangent of the pressure angle is equal to the amount of backlash.

Another machine is shown in Fig. 46. In addition to the features of the foregoing testing machine, the stationary spindle on this machine may be released from its fixed position and held against the sliding spindle by springs. When gears are mounted and revolved, the springs cause the slide to follow the mating gear. If the gears are eccentric, or noticeable errors are present in the tooth forms, this slide will move in and out. This movement is registered on the dial indicator at the end of the machine.

Another commonly used method of testing gears for concentricity consists of mounting the gears on an arbor in a lathe or on bench centers, then placing a pin between successive pairs of teeth and rolling the gear so that the pin passes by a dial indicator, see Fig. 47.

This method has the advantage of using equipment that is available in any toolroom or machine shop.

Saurer Testing Machine

One type of testing machine indicates or measures the uniformity of action between two gears. The Saurer Gear Testing Machine shown in Fig. 48 is a machine of this type.

A diagram of the operating parts is shown in Fig. 49. The gears are mounted on arbors spaced at the proper center distance. On each arbor is also mounted a plain disc, ground to the pitch diameter of the gear. On one arbor both the gear and the discs are mounted on the same sleeve. On the other, which also carries the indicating device, the gear is mounted on an external sleeve while the pitch disc is carried on the internal sleeve. The indicator is fastened to the same sleeve which carries the pitch disc, while an arm which engages with the indicator is attached to the external sleeve which carries the gear. A plate which holds the chart is mounted on the fixed arbor.

When the first spindle is revolved, the gear mounted on it drives the gear mounted on the double sleeve, while the pitch disc on the first spindle drives its mating pitch disc by friction. The indicator

Fig. 48.   Sauer Gear Testing Machine with Charting Attachment.

records any difference in the angular position of the two sleeves. If the gears are perfect, no angular movement takes place so that the indicator point remains stationary and the chart shows a smooth line.   When errors are present, the indicator point moves accord-

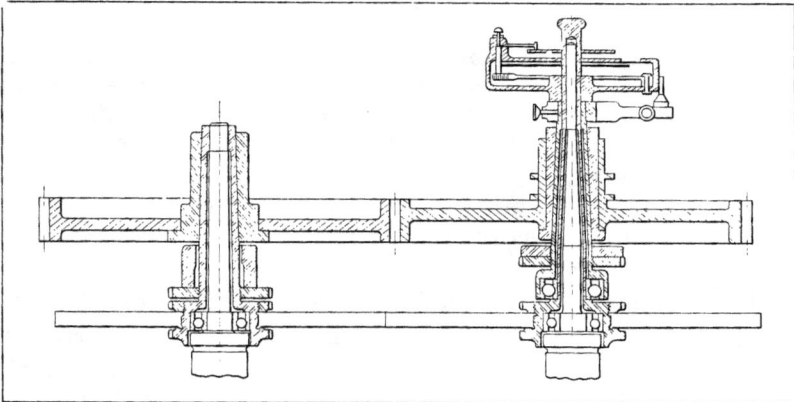

Fig. 49.   Construction of Spindles of Sauer Machine.

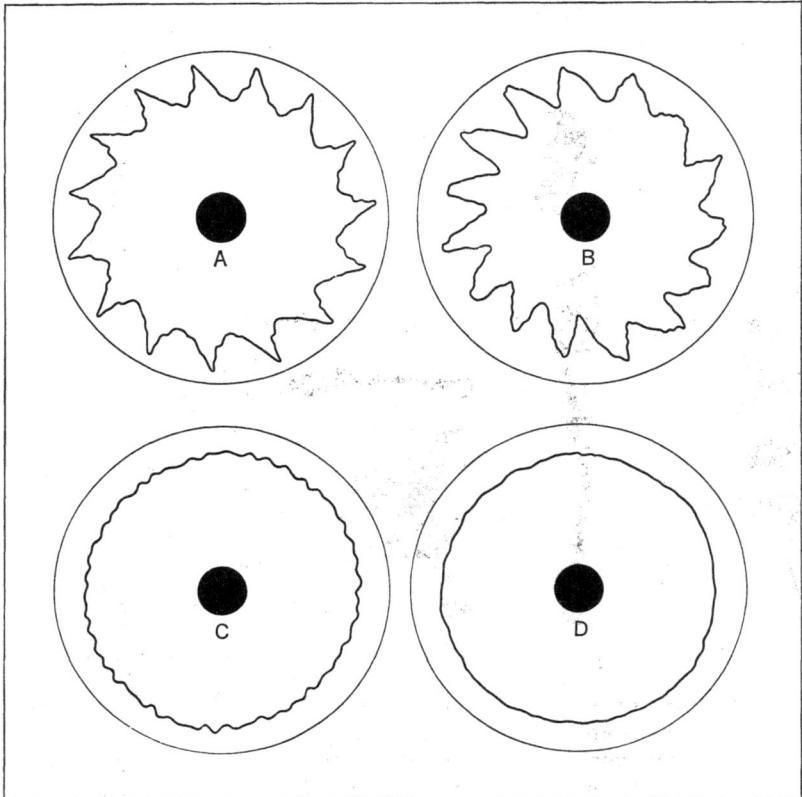

Fig. 50.   Charts of 15-tooth Gears Made on Saurer Machine.

ingly, and the chart shows a correspondingly irregular line.   If the gears are correct, and the diameters of the pitch discs correct, and no slipping takes place between them, the chart is a perfect circle.   An error in the diameter of the discs or slippage between them develops a spiral chart.

As a matter of fact, a slightly spiral chart is of advantage, as the gears may be revolved several times, and the succeeding charts closely compared.

Fig. 50 shows charts made by different pairs of 15-tooth gears produced by different methods.   The actual charts are about three inches in diameter.   A variation of about $\frac{1}{16}$ inch from a smooth line on the chart is caused by a difference of one minute in the angular position of the two sleeves on the indicator spindle. Thus an error in the gears which causes an angular variation of one minute is represented by $\frac{1}{16}$ inch on the chart. All of this error may be in one gear or it may be partly in one and partly in the other. If both are identical, a variation of $\frac{1}{16}$ inch would indicate an error of 30 seconds in each gear. By testing the gears against a tested master gear, the errors can be definitely located.

An analysis of the chart enables errors in profile, spacing, and in concentricity to be determined very closely. An eccentric gear develops an eccentric chart. If the ratio is other than one to one, it develops lobes which can be readily detected. Errors in spacing show up as steps, while errors of profile develop irregular patterns.

This machine is effective in checking pinion shaped cutters. The actual cutting edge of the cutter is in contact with the teeth of the master gear as shown in Fig. 51.

Fig. 51.   Saurer Machine Testing a Pinion-shaped Cutter.

Fig. 52.   Kavle Indicator.

Involute Testing Machine

Another type of testing machine measures the accuracy of the involute profile. These consist of a disc, which represents the base circle on which the gear is mounted, and an indicator so set up that it represents the end of the line which is unwound from the base circle to develop the involute. One instrument of this type is the Kavle Indicator shown in Fig. 52.

It consists of a baseplate, a disc of the same diameter as the base circle of the gear to be inspected, a plug or bushing to locate the gear, and a straight edge which is held in contact with the base circle disc by means of a steel ribbon 0.002 inch in thickness, and a spring. On the straight edge is mounted a 5 to 1 lever, the short arm of which is in contact with the tooth form, while the long arm is in contact with the plunger of a dial indicator. This 5 to 1 lever multiplies the reading so that each division on the dial is equivalent to 0.0002 inch. As the straight edge is rolled about the base circle, the short arm follows the tooth form of the gear. If the tooth is of true form the pointer of the indicator will not move, but if errors are present the

78

indicator will show the amount. The accuracy of the spacing of the teeth may also be inspected by the same instrument by means of the spring stop, together with the dial indicator reading.

This instrument is often used to test pinion-shaped cutters. It does not check the actual cutting edge, but checks a section in a plane perpendicular to the axis of the cutter. The reading thus found should be corrected according to the angle of the cone forming the cutting face of the cutter.

Another instrument of this type is shown in Fig. 53. It consists of a pair of straight edges on which roll two base circle discs of equal diameter. A magnetic circuit is made through the straight edge and disc to prevent slippage. It is of interest to note that when friction gears, such as a pair of discs or a disc and a straight edge are required to roll without slippage, the use of such a magnetic circuit permits considerable power to be successfully transmitted.

The gear is mounted on the shaft between the two base

Fig. 53.   Involute Testing Fixture.

circle discs. An indicator is mounted on the base of the instrument, the finger of which acts as the end of the line which is unwound from the base circle to form the involute. The operation of this instrument is identical with the one previously discussed.

The Odontometer

The Odontometer is a simple and self-contained instrument for testing the accuracy or uniformity of the gear tooth profiles and spacings of the teeth in production work. The instrument illustrated in Fig. 54 has a range of from 3 to 10 diametral pitch, may be used to check gears of any pressure angle, and can be applied to a gear while it is in place in the machine.

In effect it is composed of a section of a straight-sided rack with two parallel effective faces, one fixed and the other movable. A third face, set at an angle to the two working faces, is used to hold the fixed working face in contact with the flank of the gear tooth. The fixed registering surface is at A, the movable indicating surface at B. The third surface C holds surface A in contact with the involute sur-

Fig. 54.   Pratt & Whitney Odontometer.

face of the gear tooth. The surfaces *B* and *C* are adjustable so that gears of various pitches can be tested with the same instrument.

The indicating surface *B* is mounted on two thin flat springs *D*, which act as pivots free from backlash. The dial indicator *E* is actuated by the lever *F*, which has a ratio of 5 to 1, so that each division on the dial represents a movement of 0.0002 inch of the indicating surface *B*.

In order to explain the operation of this instrument, it is necessary to reconsider some of

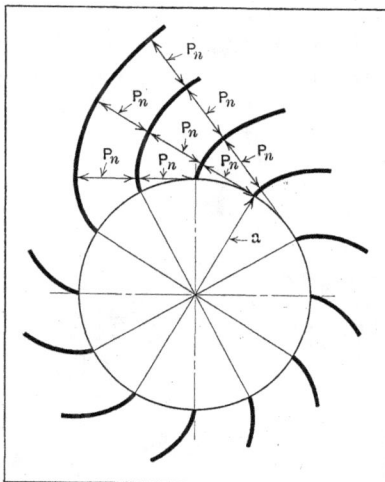

Fig. 55. Equally Spaced Involutes.

the characteristics of the involute profile. Fig. 55 shows a series of involutes equally spaced on a given base circle. The normal pitch P_n between these involutes along a line tangent to the base circle is always the same, no matter where the tangent is drawn.

Fig. 56 shows a few gear teeth and indicates the portion of the involute which the Odonometer actually tests. Note that this does not cover the entire involute surface of the gear tooth. The instrument is located over one tooth and rocked into contact with the adjacent one. When the movable face of the instrument reaches position *A*, the distance P_n should be registered, and should be constant until the instrument reaches the line *B*. Although the instrument only tests the accuracy of the involute profiles on the "overlapping" portions, these are the most essential for quiet and smooth running.

In general the instrument is used as a comparator to test the uniformity of interchangeable and mating gears. If actual measurements are required, the distance

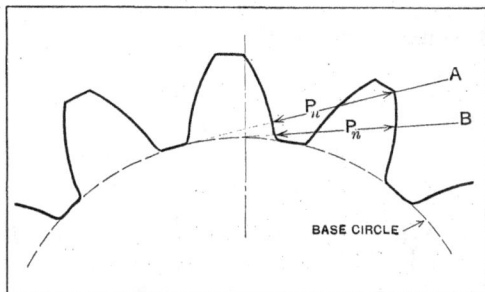

Fig. 56. Diagram Showing Action of Odontometer.

81

between the two parallel working faces of the instrument can be measured. Fig. 57 shows it set for a gear of larger pitch than in the first illustration.

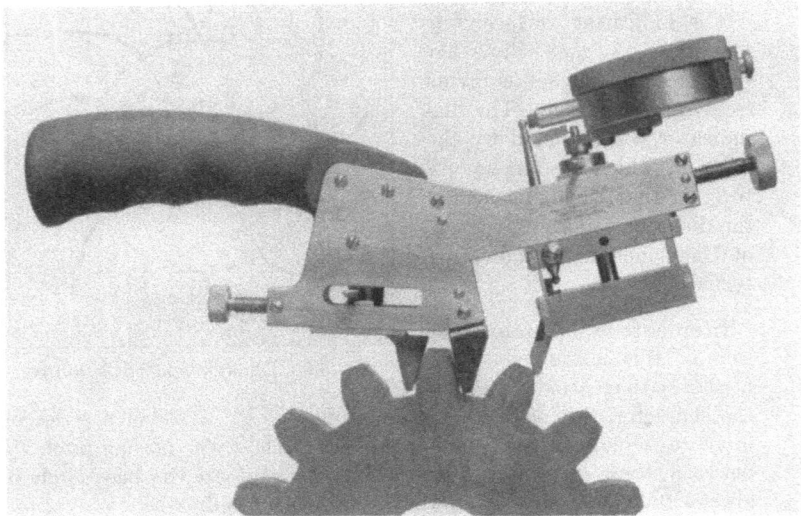

Fig. 57.   Odontometer Testing Gear of Large Pitch.

A stand is made to hold the Odontometer when gears or pinion-shaped cutters are being tested. This is illustrated in Fig. 58. In this case the gear is placed on the surface plate and rolled by the instrument. The actual cutting edge of the cutter is tested, so that errors of whatever nature are detected.

Profile Measuring Apparatus

Another type of testing instrument is shown in Fig. 59. This comprises a base, with a slide which carries the gear. A micrometer head is used to locate the top of the tooth while two dial indicators are arranged to measure the thickness. The gear is shifted from position to position, and the readings of the dial indicators recorded. After the measurements are taken, these coordinates are plotted to an enlarged scale. This type of instrument is sometimes used to get a record of the wear on the teeth of gears. A pair of gears are measured, then placed in a testing machine and run under load for a certain length of time, then removed and measured again. The successive measurements are then plotted on the same chart, which shows graphically the effects of wear.

Projection Testing Apparatus

Projection apparatus, similar to that developed for the inspection of screw threads, is often used for the inspection of

Fig. 58.   Odontometer on Stand Testing Pinion-shaped Cutter.

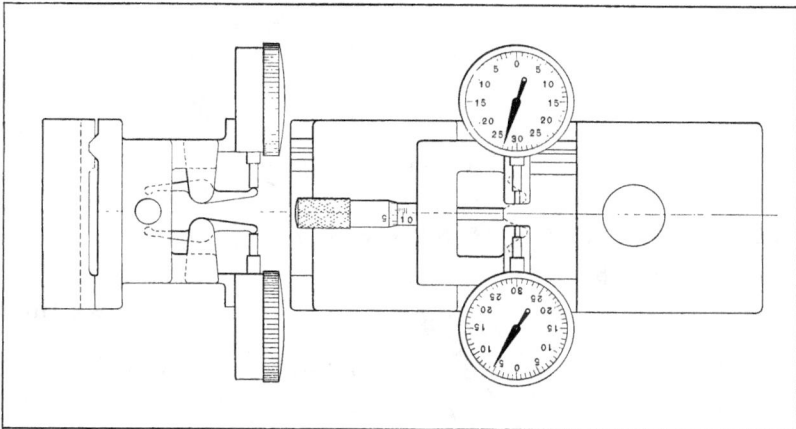

Fig. 59.   Profile Measuring Instrument.

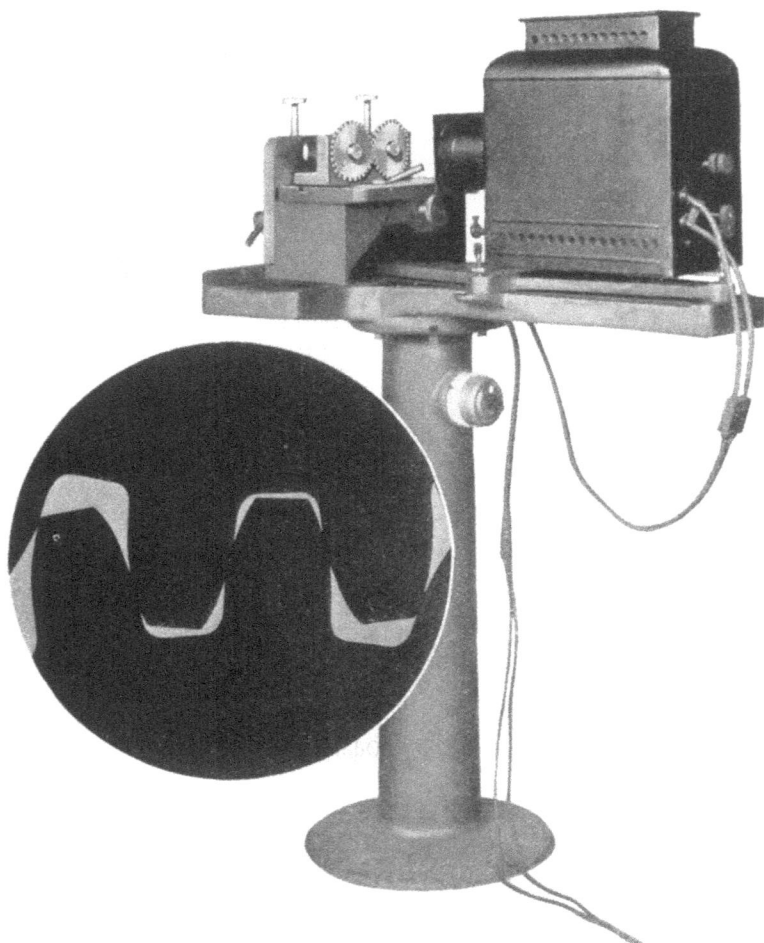

Fig. 60.   Projection Apparatus for Testing Gear Tooth Profiles.

involute profiles. One type is shown in Fig. 60. The enlarged image or shadow of the tooth profile is thrown on a screen and compared with the correct profile which is drawn there.

Another style of projection apparatus projects the image of two meshing teeth of a pair of gears, and the action between them is examined as the gears are rolled together.

Running Testing Apparatus

In order to test gears for noise, it is necessary to mount the gears and run them under load. A machine for this purpose is shown in Fig. 61. One of the spindles is mounted on a slide so that the machine can be readily set to any desired center distance. The fixed spindle is driven by a belt while the adjustable spindle is provided with a friction brake which applies the load to the gears.

Machines of this type are often provided with a dynomometer to apply the load and also to get some measure of the power transmitted. Such machines are often used when test-

Fig. 61.   Machine for Testing Gears for Quietness when Running Under Load.

ing gears to determine the nature and amount of the wear on the teeth.

CHAPTER V

Strength of Gears

Representative Formulae

LITTLE definite information is available as to the actual strength of gears, though about fifty different formulae are used for this calculation. They often give widely different results. Any of them prove reliable for gears which run at relatively low speeds and transmit a moderate amount of power; but the method of determining the strength of gears running at high speeds and under heavy loads is a problem yet to be solved.

The formulae used to determine the strength of gears may be divided into two classes. First, those which consider primarily the bending stress in the gear teeth; second, those which consider primarily the compressive stress. The actual solution must probably consider both. At present the safest plan undoubtedly is to design gears which satisfy both types of formulae. In addition, the accuracy of the gears themselves is a vital factor in determining the safe maximum load the gears can transmit, particularly when running at high pitch line velocities.

We will first consider formulae for the strength of gears which are based on the bending stresses. Of these, the best known and the one most widely used is the Lewis formula.

$$W = \text{spfy}$$

Where W = load transmitted by teeth in foot-pounds.

s = safe working stress of material in pounds per square inch.

p = circular pitch in inches.

f = width of face in inches.

y = factor depending upon the form of the tooth, whose value for different cases is given in the following table:

Number of Teeth	Factor of Strength—y		Number of Teeth	Factor of Strength—y	
	15° Involute	20° Involute		15° Involute	20° Involute
12	.067	.078	27	.100	.111
13	.070	.083	30	.102	.114
14	.072	.088	34	.104	.118
15	.075	.092	38	.107	.122
16	.077	.094	43	.110	.126
17	.080	.096	50	.112	.130
18	.083	.098	60	.114	.134
19	.087	.100	75	.116	.138
20	.090	.102	100	.118	.142
21	.092	.104	150	.120	.146
23	.094	.106	300	.122	.150
25	.097	.108	Rack	.124	.154

The factor s is determined from the elastic limit of the material and the velocity at which the gears run. A factor of safety of 2 is generally allowed for steel forgings and a larger factor of safety for castings. One equation for determining s is the Barth equation as follows:

$$s = \left(\frac{600}{600+V}\right)S.$$

Where
  $S =$ allowable static unit stress. (Elastic limit of material divided by the factor of safety.)
  $V =$ pitch line velocity in feet per minute.
  An example of a formula based on the compressive stress is the following, which has been used by an Italian firm, Luigi Pomini, Castellanza, Milan. It has been found to give good results with spur gearing and is being applied tentatively to double helical gearing:
  $P =$ load in pounds per inch of face.
  $p =$ circular pitch in inches.
  $v =$ velocity of pitch line in feet per second.
  $R =$ a factor depending on the number of teeth in the pinion and the reduction ratio.

$$P = Rp\left(\frac{1480}{v+32.8}\right)$$

For cast iron spur gears the value of P is taken as above. For steel spur gears, $3P$ is taken as the load.
  Values of R are given in the following table and apply to enclosed lubricated gearing:

Number of Teeth in Pinion	Reduction Ratio							
	1:1	1:2	1:3	1:4	1:5	1:6	1:8	1:10
12	2.8	3.4	3.8	4.2	4.36	4.54	4.8	5.0
14	3.2	3.8	4.2	4.6	4.88	5.08	5.4	5.6
16	3.5	4.2	4.64	5.06	5.36	5.58	5.84	6.1
18	3.8	4.4	5.0	5.4	5.76	5.96	6.24	6.44
20	4.2	4.9	5.4	5.9	6.2	6.4	6.88	6.9
24	5.0	5.76	6.3	6.8	7.04	7.3	7.6	7.8
28	5.7	6.4	7.04	7.6	7.88	8.14	8.5	8.64
32	6.4	7.28	7.92	8.4	8.8	9.04	9.4	...
36	7.2	8.1	8.76	9.24	9.6	9.88	...	...
40	7.9	8.84	9.56	10.28	10.44	...	...	...

*These factors have been deduced from observations on the wear of lubricated teeth with circular pitches varying from ½ inch to 2½ inches. They are valuable in determining tooth pressures for high speed gearing where wear is the determining factor.

**Another formula based on the compressive stress is the following, which uses the radius of the pitch circles as the comparative radii of curvature:

$$W = \frac{\sin 2\,a}{0.7}\left(\frac{1}{e} + \frac{1}{E}\right)\frac{C^2 + r}{1 + \dfrac{r}{R}}$$

Where

W = maximum safe tangential force as regards the compressive stress at the pitch diameters, assuming that only one pair of teeth is in mesh at a time.

a = pressure angle in degrees;

e and E = moduli of elastictiy of materials in contact in pounds per square inch. (30×10^6 lbs. per sq. in. average value of steel.)

C = maximum allowable compressive stress on the tooth face at the pitch line in pounds per square inch for each particular speed V.

f = width of face in inches.

r = radius of pitch circle of pinion.

R = radius of pitch circle of gear.

V = pitch line velocity in feet per minute.

*(See lecture by Mr. Joseph Chilton on the Manufacture and Design of Toothed Gearing, delivered before the North East) Coast Institution of Engineers and Shipbuilders, Feb. 11, 1919, at New-castle-on-Tyne, England.)

**(See article on Strength of Gears, by Mr. Joseph Jandasek, in Automotive Industries, Sept. 15 and 22, 1921.)

The Increment Load

If it were possible to cut theoretically perfect teeth, the impulse delivered to the driven gear would be smooth and continuous, fulfilling ideal conditions never attained in practice. With a pair of such gears made of an indeflectable material, variations in the driven velocity would be impossible, and the calculated static load could then be carried at any desired pitch line velocity up to a safe speed based upon centrifical forces, provided the gears were in perfect balance.

Tooth action, however, is made up of acceleration and retardations due to errors in tooth form, spacing, etc. At low speeds these errors have a relatively slight effect, but at high speeds may result in an increment load of several times the applied load. The teeth, therefore, must be sufficiently strong to carry this increment in addition to the applied load.

Various formulae are used to estimate the effect of this increment load which develops as the speed is increased. Unfortunately, it is impossible at present to establish definite values for this factor. No means exist for measuring the action of gears when running at high speeds under heavy loads. Static tests only are possible, and even these are not satisfactory. Thus, at present empirical formulae are used, among which are the following:

W_s = safe static load.
W = safe working load at specific speed.
V = pitch line velocity in feet per minute.
Barth equation:

$$W = W_s \left(\frac{600}{600 + V} \right).$$

According to the Barth equation, the increment load increases proportionately to the velocity. This equation can be safely used for low velocities, but is not satisfactory for speeds over 2000 feet per minute. Mathematically, the increment load due to irregularity of motion must be proportional to the square of the pitch line velocity, and cannot be directly proportional to it. No gears are safe at extremely high speeds, thus the increment load must increase faster than the velocity.

90

Another formula for this increment load which gives values proportional to the square of the velocity is the following: (See paper by Mr. Jandasek, reference page 89.)

$$W = \frac{W_s}{1 + \left(\dfrac{V}{1000}\right)^2}.$$

Another formula is given by Mr. C. H. Logue as follows:

$$W = W_s - F$$

Where F = increment load.

The value of F is determined from the masses of the revolving bodies, the velocities at which they run, and the accelerations and retardations which are caused by the errors in the gears. This value depends upon so many varying factors that no simple equations can be given. Each case must be calculated individually.

WALCUTT BROS. CO.
NEW YORK